Building a Nation at War

HARVARD EAST ASIAN MONOGRAPHS 457

Building a Nation at War

*Transnational Knowledge Networks and the
Development of China during and after World War II*

J. Megan Greene

Published by the Harvard University Asia Center
Distributed by Harvard University Press
Cambridge (Massachusetts) and London 2022

The Harvard University Asia Center publishes a monograph series and, in coordination with the Fairbank Center for Chinese Studies, the Korea Institute, the Reischauer Institute of Japanese Studies, and other facilities and institutes, administers research projects designed to further scholarly understanding of China, Japan, Korea, Vietnam, and other Asian countries. The Center also sponsors projects addressing multidisciplinary, transnational, and regional issues in Asia.

Cataloging-in-Publication Data is on file at the Library of Congress.

ISBN 9780674278318 (cloth)

Index by Azucena Melchor

♾ Printed on acid-free paper

Last figure below indicates year of this printing
31   30   29   28   27   26   25   24   23   22

To my mother, Sue N. Greene
December 21, 1931 to August 1, 2021

Contents

Contents

Figures and Tables

Figures

Tables

Acknowledgments

Although I did not know it at the time, this book started to come into being when Rob Culp and Eddy U invited me to participate in a conference on knowledge production in modern China for which I produced a paper on state sponsorship of scientific and technical knowledge. My research for that paper led me to want to see what more I might find on this subject, and a good starting point seemed to be Joseph Needham's papers, held in the Needham Research Institute (NRI) in Cambridge, England. With the generous support of one of the NRI's Andrew W. Mellon Research Fellowships and a great deal of help from NRI librarian John Moffett, I was able to use my three months at the NRI to get a sense of the scientific and technical landscape in inland China during the war as described in considerable detail by Needham and others whose writings he had collected. By happy accident, at around the same time, my colleague Ted Wilson, a U.S. military historian, asked if I might be interested in some documents he had collected some years before on the American War Production Mission in China. These documents led me to many more on Sino-American wartime technical engagement that are housed at the Truman Presidential Library, the Franklin D. Roosevelt Presidential Library, and the National Archives in Washington, DC. I am grateful to Ted, Rob, Eddy, and John for their early assistance with and encouragement of this project. In the intervening years, I have benefited enormously from the in-person and virtual help of archivists at the Institute of Modern History, Academia Sinica, the Second Historical

Archives in Nanjing, Academia Historica in Taipei, Britain's National Archives, Cambridge University Library, the National Archives in Washington, DC, the Franklin D. Roosevelt Library, Wesleyan University, Harvard University, Columbia University, Iowa State University, the Linda Hall Library, and the Truman Library. Vickie Doll, China and Korea librarian at the University of Kansas, has helped me locate hard-to-find sources and make effective use of online resources. I would also like to thank the Institute of East Asian Studies at University of California, Berkeley for permitting me to reuse some material from my chapter "Looking Toward the Future: State Standardization and Professionalization of Science in Wartime China," published in Robert Culp, Eddy U, and Wen-hsin Yeh, eds., *Knowledge Acts in Modern China: Ideas, Institutions, and Identities.* Numerous colleagues have given me helpful suggestions and feedback on conference papers and presentations that were in one way or another linked to this project. I am indebted to Zuoyue Wang and James Lin for their comments and suggestions in such settings. I am deeply grateful for the support of friends and colleagues who have provided encouragement and helped me manage my time and organize my thinking, especially William Bowman, Sara Gregg, Liz Mac-Gonagle, and Lisa Wolf-Wendel. Finally, I would like to thank my father, Jack P. Greene, brother, Granville Greene, and especially my husband, Tony Melchor, and daughter, Azucena Melchor, for their support and encouragement. My mother, Sue N. Greene, passed away in the summer of 2021, as I was working on revisions to this manuscript. She was an avid reader, clear thinker, great intellect, careful grammarian, loving mother, and doting grandmother who made regular trips to Lawrence to help out and support Tony's and my careers. She was a pleasure to be around and a wonderful role model in all regards but especially for her humanity and treatment of others. She did not get to read this book, but she made it possible for me to write it.

Abbreviations

AWPM	American War Production Mission in China
BIAM	Beijing Institute of Agricultural Mechanization
CIA	Central Intelligence Agency
CIC	China Industrial Cooperatives
CIE	Chinese Institute of Engineers, America Section
CPB	Central Planning Bureau
CSC	China Supply Commission
CUSA	Council for U.S. Aid
CVOC	China Vegetable Oil Corporation
CWPB	Chinese War Production Board
ECA	U.S. Economic Cooperation Administration
FEA	Foreign Economic Administration
ITA	International Training Administration
JCRR	Joint Commission on Rural Reconstruction
KMT	Kuomintang
LFCC	Liquid Fuels Control Commission
MEA	Ministry of Economic Affairs
MOAF	Ministry of Agriculture and Forestry
NARB	National Agricultural Research Bureau
NBIR	National Bureau of Industrial Research

NDSM   National Defense Science Movement
NRC   National Resources Commission
NRC-NY National Resources Commission New York
SEC   Supreme Economic Council
UNRRA   United Nations Relief and Rehabilitation Administration
UTC   Universal Trading Corporation

Notes on Romanization

I have elected to use pinyin except in cases where names of people or entities have been most commonly romanized using a different form. In cases where I do not have the Chinese characters for Chinese names or where individuals are identified in the sources by a spelling that is not pinyin, I have used the spelling contained in the sources.

INTRODUCTION

A s Japan attacked and occupied eastern China in 1937 and 1938, waves of people moved inland to provinces of which they mostly had little prior knowledge. Although Qing and early Republican administrators and Chinese and Western scholars and social activists had previously undertaken studies of the human and natural ecology of parts of this region, for both the newly relocated central government and the people who traveled with it, inland China was a comparatively unknown territory that needed to be both understood and exploited in order to sustain the dramatically expanded population, support the war effort, and provide a foundation on which to rebuild China after the war.

At the start of the war, comparatively few spots in this inland region were modernized or industrialized. Roads and transportation were, in many areas, unfit for motorized vehicles. The potential for expanded agricultural productivity, particularly in the north where the land was arid, seemed limited. Underground mineral resources were only partly mapped and largely untapped. Although a comparatively high percentage of the people who migrated from eastern and southern China were well-educated, the number of people with useful technical skills appeared to most observers to be insufficient for the enormous task of realizing the region's full productive capacity. At the same time, those scientists and technicians who had migrated to the region, though not always fully employed, were frequently full of ideas and looking for outlets for those ideas. Meanwhile, Nationalist bureaucrats and politicians sought to build in these territories a platform from which to push back against the Japanese and reassert control of China. Looking to understand, exploit, and develop the resources of inland China, Chinese political leaders, bureaucrats, scientists, and technicians embarked, during the war, on an extensive set of surveying, planning, human resource training, and institution-building activities for which they engaged in both very local exploration and experimentation *and* expansion and mobilization of transnational networks of Chinese and foreign scientists and technicians as well as foreign government entities. Collectively, these efforts show how the Nationalists and those who retreated inland with them, in spite of their extremely weak position, sought to harness and control the natural and human resources available to them to promote both regional and national development.

China's wartime scientific and technical development efforts were shaped not only by the immediate context in which they occurred and the longer-term goals of the Nationalist government. They were also shaped by the aims and interests of the Nationalists' American advisers. Although American technical assistance was typically framed as being wholly focused on helping China position itself to help win the war against Japan, in fact American advice and assistance was also laying the groundwork for a more robust postwar trading relationship that would bring China more fully into a U.S.-dominated global economy. Thus, both the Nationalist government and its U.S. technical advisers had at least one eye on the future as they worked to develop inland China's industrial and agricultural potential during the war. Both also worked, albeit in somewhat different ways, to extend the control of the Nationalist state over industry and agriculture.

The Shift toward Applied Science

The war transformed the environment within which China's scientific and technical development was taking place. Academic and industrial venues for both research and practical application of knowledge were radically different in wartime than they had been in peacetime, and access to laboratory materials, scientific literature, machine parts, and raw materials was considerably more limited for those scientists and technicians located in inland China than had been the case in coastal China before the war. The task of getting a laboratory up and running, therefore, was filled with challenges, and scientists and technicians in both academic and industrial laboratories often had to find creative and inventive solutions to the material challenges they faced. Partly because of the scarcity of resources, under wartime conditions, finding solutions to immediate needs took precedence over acquisition of foundational knowledge for both scientists and technicians and Nationalist leaders and bureaucrats. As applied science and technology rose in importance, it became a point of emphasis in development planning, particularly with regard to education and training. Although an emphasis on application of scientific and technical knowledge among certain government bureaucrats and individual scientists had already been taking form in the early 1930s, the war clarified this focus, particularly with regard to questions of state-led economic development.

In the late 1930s, Nationalist technocrats and some individual scientists and technicians rapidly came to see that, particularly in a time of war, science and technology needed to serve the nation (and state), but this view had not dominated China's prewar scientific community and there lingered among many of China's scientific elite a dedication to pure science. Peter Buck, in *American Science in Modern China*, addresses in a limited way the tension between theoretical and applied science in the prewar period. China's American advisers and leading scientists in the 1910s and '20s, he shows, extolled the merits of pure science and research, and the China Foundation, a Sino-American group funded with Boxer Indemnity moneys, was, by the early 1930s, supporting science professorships at some of China's major universities, eventually becoming the leading supporter of scientific research in China. As Buck shows, however, this emphasis on basic science led scientists down an "ever narrowing 'dangerous path' toward irrelevance" because their research achievements were "set primarily in foreign rather than Chinese frameworks." Rather than applying their knowledge in ways that were responsive to China's specific and immediate needs, scientists operated in conversation with the international scientific world.[1] In doing so, they failed to address China's immediate needs. Advocates of basic science argued, however, that they were laying the foundation for both technological development and the training of future scientists and technicians in China. In other words, basic science was, in their minds, a crucial and fundamental first step on a long-term path to development of science in modern China. In this same vein, much of British scientist Joseph Needham's work in China during the war was directed at assisting Chinese scientists to continue their basic research and maintain their connections with the international scientific world (though Needham was also quite taken with the practical work he saw some Chinese technicians doing). Furthermore, training curricula for engineers and other types of technicians also tended to emphasize the theoretical over the practical.[2] This disconnect between scientists, technicians, and real-world problems was noted by many observers during the war, who frequently commented that many of China's engineers were "deficient in practical experience" and unwilling "to 'get their hands dirty' in demonstrating practical operations."[3]

But as Eugenia Lean has shown, there were also lay scientists and entrepreneurs in early to mid-twentieth-century China who most certainly did get their hands dirty. These figures engaged in "manufacturing and production practices that fell outside of formal industrialization or formal 'science.'"[4] Using the example of Chen Diexian, who was both a poet and an industrialist, Lean shows us the emergence of what she calls a "vernacular industrialism" in which "tinkering," or a kind of amateur experimentation, was a central feature. Chen and others like him participated in the scientific and technical world without much formal training, and through practice they made significant contributions to China's technological advancement. In fact, many trained scientists and technicians were also deeply excited by the chance to make their knowledge relevant to real-world problems and the circumstances of the war gave them opportunities to do so. For example, the physiologist and biochemist Tang Peisong sounded thrilled by his early wartime activities on the scientific staff of a medical school in Guiyang, writing, "I designed everything, from a three-legged stool . . . to the actually hand-made pneumothorax machine which was rigged up from parts gotten from junk shops all by myself." Tang felt lucky to be seeing action as a scientist, even if it was not precisely the kind of action he had trained for in his British graduate program. As he wrote to his graduate school mentor, "I can sincerely say that we are having an experience that you all may envy, for as in a football game, we are the players, and you are the spectators."[5]

Chin Yueh-lin, a philosopher from National Southwest Associated University, and Wu Ching-chao, a Nationalist government economist, provided insight into the complex dimensions of wartime conversations about the role of science and technology in China at a 1943 forum at the University of Chicago. Speaking to the American audience, Chin observed that at the very beginning of the Sino-Japanese War, a group of engineering professors fled Beijing to Nanjing where they offered their services to the Nationalist government's Office of Ordnance but were rejected. Chin's co-panelist Wu Ching-chao then qualified the story by pointing out that the Office of Ordnance had been wholly composed of German-trained engineers, used to German ways of doing things and a German scientific vocabulary, whereas the engineers fleeing Beijing were American trained, and simply would not have fit in. To the American audience, this story was surprising. "I take it from what you say," one remarked, "that the government is not using the universities as we are doing in this country—to train personnel for specific jobs in the war?" Chin and Wu answered that no, with the exception of medical students and students of foreign languages, China was not using its universities in that way. But at the same time, they also observed that Chinese students were flocking to practical subjects such as engineering and that the Ministry of Education was discouraging the study of pure science, both trends that Chin found very distressing.[6]

This story illustrates several points. First, from Chin's perspective, it appeared that the government was not taking advantage of the good engineers that it had, men who were actually offering their services to the state. Second, from Wu's perspective, the engineers in Chin's example would not have been useful to the department they wished to serve. There was a disjuncture between their training and the state's needs. Third, Chin was baffled that so many students were opting to study applied sciences and that the government was encouraging this shift to the point of discouraging the study of pure sciences. Chin clearly believed, as many Chinese academics did, that the study of pure science was of paramount importance and that to discourage such studies was foolhardy, whereas Wu believed that government ministries worked best when their employees had comparable and appropriate technical training. Finally, the American audience found the entire story to be illustrative of the Nationalist government's failure to effectively marshal its resources to support the war effort.

Chin's comments reveal precisely the kind of divergence between the interests of the state and those of individual scientists that James Reardon-Anderson described in his chapter on science in Nationalist China during the war. Many scientists, Reardon-Anderson argues, were concerned about government neglect of pure research and its tendency to view science as a tool rather than a set of fields of study. At the same time, "adaptation to life in the Great Rear forced the scientists to become more self-reliant and innovative and to recognize the need to apply their knowledge to practical ends." But, he observed, "many scientists retained a sense of their professional autonomy and a commitment to learning that transcended the demands of the moment."[7] In some cases, scientists also sought to preserve their autonomy as a way of pushing back against what they perceived as excessive state interference in academia. The result of these sentiments was that academic science continued, in many fields, to focus on the theoretical rather than the practical.

What Chin seems not to have grasped was the fundamental issue that the shifting educational emphasis that he complained about was intended to address. Although, in the specific example Chin provided, the state appeared to be rejecting highly trained technical expertise, in fact, the gravitation of students toward applied sciences was not accidental. From the state's perspective, China needed technicians with specific practical skills, but the Chinese academy was not well equipped to produce such technicians. Over the course of the war, the Nationalist government worked to reform the educational system so that it would produce the kinds of skilled technicians that China so desperately needed. However, it simultaneously sought other training opportunities, especially for technicians who had already completed their formal education, and worked with both Chinese Americans and the U.S. government to create them.

Transnational Engagement and the Transmission of Knowledge

After the United States' entry into the war, heightened U.S. interest in China's military success coupled with hope on the part of the Nationalist government that the United States would help it solve its development problems led to an increase in U.S. technical assistance to China. As Ying Jia Tan has shown, before the war the Nationalist government actively solicited technical advice from and constructed technical transfer agreements with foreign firms specializing in certain high priority fields, but the scope of such activities was limited as was the Chinese capacity to negotiate such agreements on wholly beneficial terms.[8] The war both expanded that scale and in some ways altered the balance of power between Chinese technicians and institutions and their advisers in the United States. In other words, the war created a new situation in which Chinese and American bureaucrats and technicians worked together to improve China's productive capacity and orient its scientific and technical community toward finding solutions to concrete, applied problems. Chinese and American diplomats, bureaucrats, scientists, technicians, businessmen, and others did this by developing new institutions and structures that could facilitate the transfer of knowledge, skills, and technology from the United States to China.

The creation of structures through which knowledge could flow did not guarantee that the knowledge itself would be appropriate for the tasks at hand or that it would be used in the ways that the purveyors of that knowledge anticipated. The pathways for knowledge transfer that this book describes, though collaboratively constructed, generally assumed a one-way transmission of knowledge, skills, and technology from the United States to China. U.S. participants in these activities typically operated on the assumption that their advice on how to use that knowledge would be followed. These pathways were being developed at a moment when U.S. scientists, technicians, industrialists, and bureaucrats had a high level of belief in their own knowledge and methods, and confidence that these methods were both markers of American progress and superiority and a rationale for the exportation of American ways of doing things to other parts of the world. For early to mid-twentieth-century Americans and Europeans, technological development was an indicator, as Jill Lepore has observed, of civilization, and machines were "the yardstick by which Americans and Europeans measured the rest of the world."[9] If the most advanced societies in the world were advanced because of their technological superiority, then logic suggested that other parts of the world that adopted those technologies would advance, too.

Certainly, the expectations of many Chinese adopters of Western science and technology coincided with this way of thinking, as did the analytical frameworks of scholars who have described those processes. Much of the existing English-language literature on the development of modern science and technology in China focuses on efforts to bring that knowledge to China and on the challenges of that process.[10] As Grace Shen has observed, however, that scholarship focuses primarily on access, comprehension, transmission, and reception, but little of it considers the question of "why in the world a Chinese individual in the late nineteenth or early twentieth century would ever care about science," and in failing to do so, that literature simply perpetuates the assumption that "science as laid out in the West was an ineluctable good of stable content and universal appeal."[11]

In the wartime era, the answer to Shen's question seems comparatively straightforward. Modern science and technology, properly applied, could improve the situations of the Chinese people, win the war, and rebuild the nation. Although individual actors were also motivated by their own particular goals, including sheer interest, self-improvement, and the quest for employment, this was the broad framework within which most Chinese technicians and bureaucrats articulated their motivations as they sought to acquire, invent, or apply new knowledge during the war. China's international partners, on the other hand, had a narrower set of motivations that were guided by their own particular interests. They did not live in China, were not suffering the same deprivations, and had different interests in China's immediate situation and long-term survival. In fact, the expectations of the diverse actors in this story varied considerably, and those expectations were often shaped by divergent motivations.

Chinese, Chinese American, and American participants in wartime knowledge transfer activities were thus motivated by a wide range of interests and goals. At the most superficial level, they all at least claimed to be motivated by a common interest in defeating the Japanese and ending the war. But that was never the whole story. American diplomats and military leaders tended to emphasize the speedy end of the war as the primary goal of the programs they participated in. American scientists and technicians sent by the U.S. government to China often saw their role in an altruistic light, thinking of themselves as generously extending civilization and knowledge to China, though some seemed more motivated by the sheer excitement of the work and the exhilaration of being in the field and finding solutions to the scientific and technical problems they faced. Many of these actors found that Chinese responses to their efforts were sometimes unexpected, and as they tried to understand these responses, they developed their own explanations for Chinese behaviors that they found to be problematic or even mystifying.

The dominant narrative developed by American diplomats, military leaders, politicians, and industrial consultants suggested an odd combination of passivity and manipulativeness on the part of the Nationalist government and an effete inability to engage with the practical on the part of Chinese elites, including bureaucrats.[12] According to this narrative, Nationalist leaders were primarily motivated by their desire to retain power and their single-minded focus on eradicating the Communists. They had done a poor job of fending off Japanese aggression, and once the United States entered the war, they were happy to sit back and let the United States finance and manage the war effort, all the while conserving their own resources, and skimming off U.S. resources, for the postwar period.[13] At the same time, Nationalist leaders were described as being stubbornly attached to their own ways of doing things and unwilling to change. These American observers saw Chinese leaders, bureaucrats, and technicians as uninterested in innovation, hampered by excessive deference to traditional values and hierarchies, and as noted above, unwilling to get their hands dirty. These common tropes passed around within the community of American actors in China and shaped the approaches to the dissemination of knowledge to China that many of these actors took. But these views were not universal. Many U.S. observers, particularly scientific and technical experts who spent time in China during the war, met and worked with Chinese bureaucrats, scientists, and technicians whom they admired for their commitment to the nation, their skills in their fields, and their creative and innovative problem-solving efforts. Among the U.S. personnel who worked with the Nationalist government, its military, and its technocrats, therefore, there were divergent perceptions of Nationalist China and of the individual Chinese with whom they were working.

After 1941, the overriding U.S. interest in East Asia was to defeat Japan. To accomplish that goal, the U.S. government sought to strengthen the Chinese military and help China develop its war production capacity. Nationalist government officials were also eager to defeat Japan and saw U.S. assistance as essential to accomplishing that goal. But for them, the defeat of Japan was neither the end game nor even the only game; it was simply a very important step toward a much longer future that was filled with challenges. After all, both during the war and once the Japanese had retreated, the Nationalists still had to govern the nation and serve the people. For this reason, they did not always agree with the approaches that their U.S. advisers suggested, and they sometimes gently subverted them by paying lip service but then simply failing to follow the advice. At the same time, Nationalist bureaucrats and friends also sought to build new extragovernmental pathways through which they might acquire knowledge and resources that could assist them to reach all of their goals, not solely those goals of which the U.S. government approved. This process got under way in the early 1940s but accelerated as the war came to a close. These activities may well have contributed to the poor opinion of the Nationalists that many Americans developed, but it is important to recognize in them that the Nationalists were seeking not just to win the war but also to position themselves to build a strong postwar China. The war provided the Nationalist government with opportunities to mobilize resources at home and abroad that would serve *both* its wartime and postwar interests, and they would have been foolish not to take advantage of such opportunities.

It turns out that in spite of the criticisms of their American contemporaries, the Nationalists were not the only participants in these networks who were at least as concerned about what would happen after the war as about the progression of the war itself. U.S. businesses and the technical advisers they sent to China as part of U.S. government missions were also motivated by postwar goals. In the United States, numerous businesses opened their doors to Chinese trainees, hired Chinese students stranded in the United States, and permitted their own engineers and technicians to travel to China to work for U.S. government-led technical assistance programs. They did so in part because they were asked to by their government, but also because it was in their own interest. When hiring Chinese students, for example, they saw an opportunity to staff their operations at a time when many American men were being called up for military service. By establishing relationships with Chinese government trainees who would return to China to work in government agencies and by exposing trainees and students to their own products as well as their ways of doing things, they sought to shape Chinese industrial expectations, norms, and standards in ways that they hoped would open the postwar Chinese market to their products. Many American participants in these programs were just as motivated by the promise of postwar opportunity as were the Chinese participants.

The war provided opportunities, or seemed to provide them, to most of the actors participating in the scientific and technical activities described in this book. In China, individual Chinese scientists and technicians, for example, engaged in activities through which they simultaneously served national needs and positioned themselves for postwar success. New landscapes served as laboratories for Chinese scientists cataloging flora and fauna and describing the human geography and economic output of the region. Chinese government planners used the war period to undertake systematic studies of China's western regions and develop data-driven plans for postwar economic development based on what they found there and what they knew of the territories from which they had fled. And the Nationalist government also took advantage of opportunities to extend their authority over parts of China where they previously had limited presence. For many American participants, there was the promise of postwar business and financial gain, but the wartime and immediate postwar scientific and technical interactions between China and the United States also fed into an American sense of technical superiority and spoke to an American desire to extend its knowledge to other parts of the world.

In the view of Americans engaged in the international dissemination of technical knowledge, America was exporting more than just technology—it was exporting a whole system of thought and action that surrounded that technology. When discussing their activities with and in China, these figures frequently described what they had to offer as "know-how."[14] Know-how, as they defined it, was quintessentially American, a combination of superior scientific and technical knowledge, work ethic, and a "can-do" attitude that they believed characterized the American entrepreneurial spirit. But just how applicable was American know-how to other contexts? What would result from the application of American know-how to China? Many American scientific and technical advisers to China recognized that the context for the application of knowledge in China was different than it was in the United States, but they did not necessarily recognize that these differences might have an impact on how American know-how could be applied in the Chinese case. U.S. engineers, in particular, tended to talk about know-how as if it were universal, ignoring historical contingencies as they did so. They often failed to recognize that implementation of American know-how was dependent on access to resources and engagement with an American-style system of production, supply, and infrastructure.

As Leo Marx has effectively shown, technologies have frequently led to the creation of entire sociotechnological systems that are required for the smooth functioning of those technologies. This process can lead eventually to a "blurring of the boundary between the material (physical, or artifactual) components of these large socio-technological systems and the other, bureaucratic and ideological components" as well as an erosion of the boundaries "separating the whole technological system from the surrounding society and culture."[15] But as David Edgerton's use-based approach to the study of technology shows, technologies do get divorced from the sociotechnological systems in which they are created, and in new contexts, as creole technologies, they may take on new functions and generate entirely different systems.[16]

Chinese circumstances during the war were markedly different from those in the United States at the same time. Technologies could not always be readily transferred from the United States to China. Not only was the broader economic and infrastructural context different, but the bureaucratic and ideological components Marx refers to were also distinct. Moreover, the participants in these technology transfer activities often did not share the same set of assumptions about what kinds of technological change were possible or necessary in China, or what the goals of such change should be. Those who participated in the knowledge networks that this book describes often struggled with how to implement American know-how in a non-U.S. context and their writings often revealed the complexities of these processes.[17]

In fact, the circulation of knowledge is not typically a straightforward process. It is dependent on go-betweens, who mediate interactions across frontiers and between cultures and through whom knowledge is translated, transformed, and reconfigured.[18] Whereas much scholarship looking at the adoption of Western knowledge in China has taken a diffusionist approach that emphasizes the outward spread of knowledge "from a single point, or 'centre' to the 'periphery,'" this book shows processes that could better be described as circulation, as defined by Kapil Raj.[19]

Many of the American partners in the knowledge networks described in this book appear to have imagined that they were participating in a top-down diffusion of knowledge through which their knowledge and know-how would be imparted to receptive Chinese technicians who would implement it directly as it had been received. Certainly, their descriptions of their activities along with their complaints about Chinese failure to participate on American terms indicate that in their view, the United States had both superior technical capacity and the upper hand in the scientific and technical relationship. Moreover, because it was providing support for many of these activities, the U.S. government actively sought to control them. For example, the U.S. embassy in Chongqing vetted Nationalist government requests for technical experts and supplied experts only in the fields that embassy personnel deemed to have merit. Later on, when the U.S. government began to pour more resources into building up China's military-industrial capacity, the Roosevelt administration determined that the only way the Nationalist government could be trusted to do this well was if it created a U.S.-style administrative unit to oversee the program.

Chinese politicians, bureaucrats, scientists, and technicians, however, were not simply vessels to receive American knowledge. They were active agents in the process of knowledge transmission and in the production of creole technologies and systems of technological organization. They had particular needs and interests and sought knowledge to meet those needs. Their own goals did not always conform to the goals their American partners told them they should have. They made formal requests for assistance to the U.S. government, but they also mobilized independent, non-government actors to help them acquire knowledge and technologies. When participating in training programs in the United States, individual Chinese technicians were also regularly engaged in a process of sifting through information to figure out what they could actually use in their own home contexts. And when working on specific technical projects in China, they adapted technologies to the material and infrastructural conditions in which they were working as well as to their own specific goals for those projects. No matter what the hopes of the American partners might have been, and despite the fact that the United States was clearly the dominant partner in the relationship, the United States could not control the way the knowledge it shared would be used, nor could it even control the kind of knowledge that the Nationalists were acquiring.

The Nationalist government had a very specific set of goals for mobilizing transnational networks. They hoped that new pathways for the training of applied scientists and technicians would help them address China's immediate and long-term needs. But to achieve these goals, they had to manage the relationships in ways that served China's needs as China's own technocrats perceived them, rather than just adopting foreign ideas about China's development wholesale. Grace Shen has argued that in the field of geology, "the wartime experience galvanized the community [of geologists] and gave it the confidence to privilege the physical evidence of the earth over the naysaying foreign authorities."[20] China's decision to develop oil fields in Gansu even after virtually all foreign prewar and wartime advisers argued that this would not be a fruitful use of resources effectively supports Shen's argument, showing that the war empowered geologists to trust in their own knowledge and strike out on an independent path. Nonetheless, even in this field, or perhaps especially in this field, since China had such a critical need for fuel, leaders of the Nationalist government's National Resources Commission recognized that they did still need foreign advice and worked with allies in the United States to set up training opportunities for the petroleum engineers working in the oil field.[21] So although necessity and isolation did fuel a kind of self-confidence that led Chinese scientists and technicians to take risks and find creative solutions to immediate problems, the Nationalist government increasingly sought foreign technical assistance in the scientific and technical realms it prioritized. As it did so, however, it sought in various ways to manage that assistance so that it would serve Chinese, rather than American, ends. In particular, the Nationalist government increasingly acted on the belief that one of the biggest obstacles to China's development was a shortage of well-trained technical personnel, and that in the short run, at any rate, many of those people needed to be trained by Americans.

American training of Chinese students was not new to the wartime period. Such activities had been going on for some time both in the United States and in China. Numerous studies of the export of American biomedicine to China show us the extent to which entities like the Rockefeller Foundation shaped Chinese medical education and practice in the early twentieth century, for example.[22] Study abroad was also common for graduate students. Although in some scientific and technical fields such as geology, Chinese students were more likely to pursue graduate education in Europe, there were strong pipelines of students in some fields, such as agriculture, physics, and economics to U.S. institutions of higher education.[23] Once trained, Chinese students abroad typically returned to take up university or government positions in China where they trained Chinese students in ways that were grounded in their own educational experiences. Some Chinese students in the United States had the chance to pursue postgraduate practical training in the United States, but in general, opportunities for practical training either abroad or in China were not abundant.[24] An important innovation during the war was the development of large-scale practical training programs for professionals already working in the field.

Such programs emerged with the aid of Chinese Americans, Chinese living in America, friends of China, the U.S. government, and representatives of the Nationalist government in the United States. Although existing scholarship tells us something about academic networks, the activities of some friends of China, and certain dimensions of the state-to-state relationship between the United States and China, there is very little scholarly work that provides insight into the role played by Chinese Americans and Chinese in America in China's nation-building efforts during the Sino-Japanese War.[25] One of the few such works, Kevin Scott Wong's *Americans First: Chinese-Americans and the Second World War*, focuses primarily on the military and domestic industrial contributions made by Chinese Americans during the war.[26] Madeline Hsu examines the experiences in America of student refugees who were unable to return to China during the war.[27] Existing scholarship tells us little, however, about well-educated and well-connected Chinese Americans and Chinese in America who played important roles as advocates for Chinese interests and also sought other ways to contribute to China's war effort, and that is one of the scholarly contributions this book makes.

Both Chinese in America and Chinese Americans worked to facilitate scientific and technical engagement between the two countries, often as volunteers. These individuals, who were not always directly connected to either government, built structures within which Chinese in America could acquire training for later use in China and that transmitted technical information and knowledge back to China and coordinated groups of Chinese scientists and technicians in the United States. Some, though not all, of these structures were supported by the Nationalist and U.S. governments, and most of them included technicians associated with the Nationalist government among their members. Through some of these structures, Chinese in America developed relationships with individual American entrepreneurs and engineers, relationships that in some cases led to transmission of knowledge that fell outside the scope of the specific fields the U.S. government was emphasizing in its own direct programs for knowledge transfer to China.

Much American technical advice came to China within the context of programs that were funded by and constructed under the direction of the U.S. government. A good deal of scholarship exists that describes the experience of American military advisers in China during the wartime period; but again, there is a scarcity of material on nonmilitary technical assistance. Wilma Fairbank's semischolarly book on the U.S. State Department's cultural relations program provides the most comprehensive study of a wartime technical assistance program, one that she herself worked on. The only study of the American War Production Mission in China was produced in early 1946, immediately following the close of the mission, by Mabel Gragg, a member of Harry Truman's White House staff.[28] These programs, though quite distinct from each other, both aimed to bring American technical expertise to bear on China's immediate technical problems. In addition, they both led to the creation of networks, patterns of engagement, and in the latter case, institutions that served as models for continued technical relations in the years following the war.

The War as a Defining Moment

Perhaps because Nationalist-controlled China suffered from such terrible inflation both during and after the war and because the counternarrative of Chinese Communist success in the 1940s has given such weight to the Nationalists' failure to provide economic leadership, scholars have tended to overlook state-led development efforts undertaken by the Nationalists in inland China during the war, though there is some work that recognizes the wartime as having built foundations that were later significant for either the Nationalists or the Chinese Communist Party (CCP). Early work on this topic includes William Kirby's article on "Continuity and Change in Modern China" in which he argues that there was "significant policy and personnel continuity" between the Nationalist and CCP eras. Joseph Esherick similarly observes that there were "areas where the Guomindang paved the way for the Communists, where the latter built on the foundations laid by the former."[29] More recently, Morris Bian has taken the lead in developing this field, arguing that the Nationalists' wartime efforts to rationalize government planning formed a foundation for the later emphasis on central planning and planned economies that characterized both the Nationalists and the CCP after 1949.[30] Bian has also argued that the transformations in regional state enterprise in Guizhou during and after the war show that "the impact of the Chinese Revolution on Guizhou's regional economic institutions was far less revolutionary than what we have been led to believe."[31] In other words, the Sino-Japanese War period was not only an important transformational time with regard to the development of twentieth-century state-owned enterprise; it was also a period during which patterns of economic management emerged that continued to develop not only immediately after the war but after the founding of the People's Republic of China as well. Judd Kinzley looks comparatively at two moments, one during the war and one in the 1960s, in which he finds continuity in the reactions of two separate governments to external crises. Kinzley's article on the development of Sichuan's Panzhihua from a village into a major steel-producing area reveals that external crises (the Sino-Japanese War in the 1930s and tensions with the USSR and the United States in the 1960s) were the driving force behind the economic development policies of two different governments at two different moments. Wars with external actors catalyzed the development of Panzhihua's steel industry and did so in similar ways under two very different political systems.[32] Ying Jia Tan has shown the importance of war in shaping the development of the power industry in China, and also shows concrete ways in which the long-term development of that industry was built on foundations that were established during the Sino-Japanese War.[33] Like these works, this book also aims to explore continuities across the temporal dividing lines that we have used to carve up our narrative of modern China. It does so by looking at the ways in which wartime modes of planning, training, and transnational technical engagement served as foundations for the postwar period in China and Taiwan. In particular, it shows that the human relationships that were developed within the framework of wartime technical exchange and patterns of institutional interaction that characterized that exchange continued to be significant in both China and Taiwan in the years immediately following the war and after 1949.

This book also shows how the war created opportunities for the Nationalist state as well as individual Chinese scientists and technicians. Studies of the United States during World War II have shown how both government and other institutions were able to use the war for their own benefits. For example, Christophe Lecuyer has shown how the war accelerated processes that were already under way at the Massachusetts Institute of Technology (MIT) and provided a context in which MIT could both promote and benefit from the establishment of a new system of federal patronage for scientific and technical research in nongovernment institutions. Of course, the dramatically increased federal patronage that benefited MIT also brought that institution more squarely into the state's orbit and put its human resources at the disposal of the state.[34] The Nationalist government also drew some Chinese academics into government service during the war, and those academics, like the MIT faculty and administrators, had opportunities to advocate for the projects they felt were most essential to the war effort, though the scope of patronage was considerably more limited.

On a somewhat different track, Susan Lindee has observed that "everything humans know about nature can become a resource for state power" and has written about how scientists have participated in the production of a plethora of ways to damage people and helped in the development of a marriage between science, technology, the scientific method, and warfare that has served the interests of states.[35] At the same time, she has observed that wars have served as important training grounds for scientists. As this book will show, the war most certainly operated as a training ground for Chinese scientists and technicians in both deliberate and unintended ways. In addition to the planned training activities described in this book, the war created an environment that compelled scientists and technicians to retool and adapt their knowledge to their circumstances, which they often did through a process of tinkering, to borrow Lean's terminology. The war also knitted scientists and technicians to the state through new institutions, training programs, and research projects on the one hand and common goals on the other hand.

Perhaps even more central to this relationship than the technologies of war and killing that Lindee focuses on, however, were scientific and technical innovations related to a wider-scale economic modernization. Certainly, projects aimed at increasing the production of fuels, improvement of infrastructure, and manufacture of weapons, airplanes, and trucks all had direct military application. But many of those projects would also serve nonmilitary needs and many of the people engaged in those efforts were at least as interested in those applications as in the military ones. For example, infrastructural development would support economic integration as well as the transportation of nonmilitary goods and people. Expanded agricultural production would not only help feed the army, but it would also help feed the general population. So although the war gave the state opportunities to exert greater control over the scientific and technical community, the Nationalist state did not, by and large, exploit those opportunities to build more innovative killing machines. The interplay between war, science and technology, and the state in the Chinese context thus yielded somewhat different results than in the Euro-American context that Lindee describes.

This book's contribution sits at the nexus of the broad themes I have explored here. It describes the ways in which the Nationalist government worked with partners in the United States during and after the war to develop China's productive capacity, arguing that the war led to fundamental changes in how the Nationalist government both promoted and sought to control and utilize science and technology. In particular, the war catalyzed the Nationalist government's shift in emphasis from pure to applied sciences, it provided opportunities for comprehensive economic planning grounded in scientific study and data collection, it stimulated a strong new focus on human resource development, and it created opportunities for China to mobilize transnational networks (particularly Sino-American networks) to aid in its wartime "national reconstruction." This book also shows how the Nationalist government sought agency in its engagement with the United States, even from a position of weakness, and it explores the motivations of both public and private American and Chinese participants in wartime and postwar knowledge networks, arguing that although the Nationalists were clearly focused on using those networks to support their postwar development plans, Americans were also highly motivated by opportunities for postwar gain. The book further argues that these wartime approaches to conceptualizing and addressing China's needs served as foundations for continuing efforts at scientific and technical development in the postwar period in both mainland China and Taiwan. Although the wartime institutions that were constructed to manage this knowledge transfer were not maintained after the war, the networks that both created and emerged from those institutions continued to function, and through them, knowledge continued to flow into China, at least until 1949, and Taiwan.

Chapter 1 explores ways in which a variety of Chinese actors developed plans to harness and expand the productive capacity of inland China. Perhaps because they found themselves operating in underdeveloped territories that suddenly needed to provide for a growing population, economic development was very much on the minds of many Chinese scientists, technicians, and bureaucrats who had migrated inland during the war. Consequently, both because of and in spite of the war, many of these people found themselves using their time during the war to build plans for both regional and national development. To this end, they undertook expeditions and surveys of townships, counties, and provinces to collect knowledge that would form the basis of both small- and large-scale development plans. The war provided Chinese bureaucrats, technocrats, and academicians with opportunities to undertake systematic studies and develop data-driven plans for development, and through the acquisition of knowledge and the development of new plans to extend the authority of the Nationalist government over the hinterland.

Many of these surveyors and planners recognized that implementation of development plans would require personnel, and in particular, well-trained scientific and technical personnel with skills that could be applied to the problems at hand. In an attempt to meet those needs, the Nationalist government undertook revisions of the educational system and promoted mass science education. After the American entry into the war, the Nationalist government requested that the U.S. State Department send scientific and technical experts to China to provide on-the-ground training and technical advice. The unfolding of this program reveals some of the underlying tensions in the relationship between the Nationalist government and the U.S. government and also shows how the different actors were motivated by different goals. These activities are the subject of chapter 2.

Chapter 3 focuses on a separate set of human resource development strategies that took place in the United States. It examines structures that were developed during the war by both government and nongovernment actors to facilitate the training in the United States of Chinese students and technicians in fields deemed important by the Nationalist government. It highlights the important role played by Chinese Americans and Chinese in America in the construction of structures that helped trainees become part of a transnational knowledge network and shows how those structures also aimed to ensure that trainees understood themselves to be participating in a broad nation-building enterprise.

By 1944, the U.S. government had invested considerable resources in helping China develop, but U.S. military leaders and diplomats were not convinced that China was utilizing those resources to the best effect and sought to put their own imprint on Chinese development efforts with an eye to increasing what they called war production. The Chinese War Production Board, modeled on the U.S. War Production Board, and the American War Production Mission in China are the subjects of chapter 4. With their creation, the United States funneled more technical assistance and advice into China and attempted to help Chinese planners revise their approach to economic planning, particularly in the short term. Chinese planners did not always respond to U.S. advice in the way their advisers hoped, however, and sought to manage the new institutions in ways that served Chinese, rather than American, interests. Through these institutions we learn more about the complex relationship between the United States and China during the war and we see the laying of foundations that would underpin a very different kind of postwar engagement.

With the end of the war and the declining interest of the U.S. government in supporting Chinese development projects, some of the institutional apparatuses that had been constructed to facilitate the flow of knowledge between the United States and China ceased to exist and others took on new characteristics. The final chapter examines how wartime patterns of planning, training, and transnational technical exchange both continued and changed as the Nationalist government had to figure out how to govern all of China and as the U.S. government was increasingly replaced by U.S. private enterprise as partners in technical exchange.

CHAPTER I

Surveying and Planning

In the late 1920s and early 1930s, the young Nationalist government made it clear through a series of institution-building efforts that even though it was still consolidating control over Chinese territories, it was going to take a developmentalist approach to government. As the work of such scholars as Margherita Zanasi, William Kirby, and Morris Bian makes clear, the new government prioritized economic development and sought to insert the state into the economic activities of the nation.[1] Most of these efforts, however, were concentrated in or near eastern urban areas. Even though the state worked with provincial governments to set up some kinds of provincial-level institutions, the extent to which China's western provinces were integrated into these nascent systems prior to the war was limited. Despite the war interrupting state-led economic development efforts in the regions that came under Japanese control, it created opportunities for such activities in the inland regions over which the Nationalist government needed (for both immediate and long-term reasons) to assert greater control than in the past.

At the same time, the imperative of national salvation combined with the need to find employment led many of China's scientists and technicians to participate in these activities either as direct partners with the state or as independent researchers. The many active scientists, technicians, and social scientists who retreated with the Nationalist government to western China were faced with decisions about how to use their time and what to do with themselves. They now resided in a vast "new" landscape about which they knew little and they were there not because they wished to be but because the war had forced them to be. As they explored their new surroundings, it quickly became apparent that this landscape had considerable economic potential and was filled with resources that could be pressed into immediate service to support the newly enlarged population as well as the war effort.

In many regards, the Nationalist government was simply carrying on during the war with a set of activities it had embarked on prior to the war. Government organs such as the National Resources Commission (NRC) and the Ministry of Education, for example, both continued to consolidate control over industry and higher education. But increased access to China's western frontier regions also afforded the government with opportunities to organize fact-finding expeditions in the northwest and Xikang and to use the information it gathered to create plans for the development of their economies. At the same time, and especially with the establishment in 1940 of a new Central Planning Bureau (CPB), the government created new bureaucratic mechanisms that were designed to encourage the production of data-driven development plans. These processes all aided in efforts to create in China's hinterland a knowledge-gathering and planning infrastructure that would be needed to develop the region and integrate it more fully into the nation. By engaging in these activities, the Nationalist government took advantage of its move inland to further its agenda of becoming a modern technocratic state. It was assisted in these efforts by numerous scientists and technicians who, looking to further their own personal goals—to find employment, gain prestige, or claim expertise—also participated in similar, sometimes state-sponsored, knowledge collection and planning activities. For all of these actors, the path to defending, reclaiming, and rebuilding China was rooted in economic development, and members of both groups devoted their time during the war to systematic study of the economic potential of China's western regions with an eye to bringing those territories more squarely into the service of the nation.

The Nationalist government, from its 1938 base in Wuhan, promulgated a Program of Armed Resistance and National Reconstruction, the very name of which captured the essence of the challenge the Nationalists faced in trying to build an economy at the same time it was fighting a war. As banker and former minister of communications Zhang Jia'ao

張嘉璈 observed at a 1944 Chinese Institute of Engineers meeting in New York, "The general concern in China today is to Resist and Reconstruct simultaneously."[2] Reconstruction was by no means a new term that emerged during the war. Rather, it was a founding principle of the Kuomintang (KMT). Sun Yatsen 孫中山, founder of the KMT, had utilized the term to describe the rebuilding that China required as a result of its economic and social disintegration during the late Qing and the term subsequently became part of the standard vocabulary of the KMT. By 1928, a national reconstruction commission (建設委員會) had been established and provincial governments had appointed provincial commissioners of reconstruction (建設廳廳長). The National Reconstruction Commission lasted only until 1938, when its responsibilities were subsumed by the Ministry of Economic Affairs (MEA) and the NRC. However, the idea of reconstruction continued to be central to KMT planning throughout the war, and with the Nationalist government's retreat inland, reconstruction took on additional meanings. Now there were new disasters from which to recover and rebuild, as well as land, infrastructure, and industry that would need to be reclaimed and reconstructed. If reconstruction was a concept that for Sun had referred to a sense of a lost, great past, for the KMT of the wartime period, that sense of loss was all the more immediate and visceral, and the idea of reconstruction began to aim not just at immediate economic development projects but also at an imagined postwar moment when China would once again be whole and the national government would be able to embark on a rebuilding process.

Reconstruction entailed planning, but that planning could be for short- as well as long-term economic development. Although it kept one eye on the postwar future, much of the government planning that took place during the war was about the immediate future. Following in a pattern well established by the Nanjing government, government ministries, offices, and enterprises were required throughout the war, but particularly after the establishment of the CPB in 1940, to engage in annual planning exercises that focused on the goals for the year.

Interestingly, planning was not the exclusive domain of the government during the war. Other groups and individuals, looking to solve either short- or long-term problems, also drew up plans based on their knowledge and technical skills, hoping that the government would recognize their expertise and adopt their plans. In some cases, government leaders also mobilized nongovernment intellectuals and technicians to participate in planning activities that took place outside of formal state organs.

By and large, as we shall see later in this chapter, planning, even planning undertaken by government ministries, was collaborative, involving meetings, conversations, and multiple drafts commented on by a variety of stakeholders. The fact that nongovernment groups and individuals were also producing plans and engaging in discussions about planning indicates that the idea that plans were needed was generally accepted in intellectual circles. In fact, groups of Chinese scholars and technicians sometimes got together simply to discuss the very idea of planning as well as to debate potential elements of plans that the government might adopt. This idea—that the nation needed plans to guide its development—was unquestionably one that the state promoted, but it was also an idea that spoke to highly educated technical experts and academics who wished to contribute to national reconstruction but did not necessarily know how to do so.

Chinese abroad were especially eager to contribute to the planning process, understanding that it was one of a small number of ways in which they could make a difference, given that they were not on the ground in China. As the Chinese Institute of Engineers, America Section (CIE) observed in a statement on postwar industrialization, "We Chinese engineers in the United States are anxious to serve to the best of our abilities, though we cannot participate directly in the planning in China at this moment." To that end, in 1943, they held a forum on postwar industrialization with the aim of presenting "considered opinions on the various aspects of the industrialization problem in the hope that such opinions may prove helpful to the authorities in China who are entrusted with the task of planning."[3] For educated Chinese abroad as well as for many educated Chinese in China, planning *was* action. It was the only real way for them to contribute.

Surveys, Expeditions, and Data Collection

Whereas Nationalist government organs had been criticized in the Nanjing decade (1927–1937) for making plans that were disconnected from reality and thus unrealistic, costly, or otherwise fatally flawed, by the mid-1930s, with the creation of the NRC, in particular, it had begun taking a more coherent and data-driven approach to planning. By the wartime period, it was generally understood among bureaucrats as well as academics outside of government that good development plans should be grounded in knowledge.[4] To make plans, therefore, government bureaucrats and others needed to collect information, an activity in which they engaged with particular vigor in China's western provinces, areas with which most Chinese bureaucrats and technicians were largely unfamiliar prior to the war. Groups like the NRC and the Geological Survey had undertaken some prewar surveys of natural and agricultural resources of inland China that served as a foundation for some wartime planning. T. H. Shen 沈宗瀚, for example, noted that agricultural survey data he had collected in Guiyang between 1934 and 1937 for the NRC "became most useful for planning agricultural programs during the war." In addition, his observations of Guiyang's institutional dysfunction in the immediate prewar years helped him to reorganize the institutions that oversaw agricultural research and extension during the war.[5] For most Chinese bureaucrats, technocrats, and academics, however, the west was unfamiliar territory, and so the war years provided a prolonged opportunity for data collection through surveys, expeditions, and county-level studies. Newly relocated government institutions were particularly keen to learn about the resources that were now available to them. For example, the National Bureau of Industrial Research's (NBIR's) late-1938 two-year plan identified as its most immediate need a survey of resources in China's southwest and their use.[6]

Expeditions and surveys of China's western provinces were by no means new to the wartime period. The Qing government did them and so did westerners such as Aurel Stein, Joseph Rock, David Crockett Graham, and various other European and American botanists, naturalists, ethnographers, and geologists in the nineteenth and early twentieth centuries.[7] By the 1920s and early 1930s, the interest in surveying had more thoroughly permeated Chinese circles, and Western-trained Chinese scholars in fields such as agricultural economics and geology were routinely collecting data on land use, topography, mineral resources, and prices of agricultural products. Until the mid-1930s, however, the majority of these Chinese studies tended to focus on locations in coastal or central China, areas to which scientists and social scientists in the major coastal cities had comparatively easy access. Geologists went farther afield, exploring mineral deposits across China, sometimes in the company of businessmen or Western geologists under contract with Western businesses such as Standard Oil.[8] Also in the mid-1930s, Chinese geographers began surveying landscapes in China's west and southwest, and the NRC started sending agricultural experts such as T. H. Shen inland to gather data on important agricultural products such as tung oil and cotton.[9]

During the war, numerous scholars and bureaucrats surveyed a wide variety of features of China's western provinces. These surveys did not all result from any systematic or comprehensive government effort, though many of the people undertaking the surveys were employed by government organs and some were even tasked with devising provincial-, regional-, and national-scale development plans by the central government. Others, however, were academic researchers operating more or less independently. Many of these surveys aimed to collect data on and describe the state of some particular dimension of the economy, often with an eye to providing a foundation for decisions about how best to encourage its further development.

In 1938–1939, for example, W. Y. Swen and S. L. Chu of the College of Agriculture and Forestry of Jinling University undertook a massive economic study of tung oil production and marketing that compiled data from numerous sources and provided a comprehensive set of information about a major economic resource across an entire large province, Sichuan. At the time the study was initiated, tung oil was still one of China's major exports, and given that roughly one-third of China's tung oil came from Sichuan, it seemed to be one of the most promising vehicles for bringing wealth into the region. With an eye to expanding Sichuan's export economy, then, the authors collected an extensive set of data from July 1938 to June 1939 on forty major wood oil-producing areas in Sichuan, ultimately producing a 319-page report that analyzed patterns of tung tree cultivation, oil production methods, transportation and marketing of oil, and government measures regarding tung oil.[10]

By the time of publication, however, owing to the impact of the war on export routes out of western China, tung oil was no longer a major export commodity, and because of the consequent drop in price, farmers in Sichuan were not only not planting new tung trees, but they were cutting down old ones to plant food crops instead, in spite of the fact that both government-run and private enterprises were at that time collecting tung oil for refinement into fuel.[11] These shifting conditions meant that almost by the time it was completed (and well before it was actually published), the tung oil survey did not have the same import or relevance that its authors had originally imagined it would have. Nonetheless, the survey provides an enormous amount of data on one important facet of Sichuan's agricultural economy and it was later used as a source for the *Economical Atlas of Sichuan*, a publication that aimed to provide a comprehensive set of economic data about the province that would be useful to bureaucrats planning for Sichuan's postwar redevelopment.[12]

In addition to large-scale, province-wide undertakings such as the tung oil survey, numerous scholars performed small-scale studies of specific regions throughout the southwest. For example, *Acta Geographica Sinica* (地理學報) published a number of county-level land-use surveys in the mid-1940s. One such survey of Chenggong County in Yunnan, an area twenty kilometers south of Kunming, described in considerable detail land-use patterns and agricultural output as well as the transportation networks and market places where agricultural products were sold. The study's value rested not only in its documentation of local agricultural and economic practices but also in its recommendations for possible future change in those practices: "As cultivation of fruits is more profitable than other dry crops, it is suggested that dry fields should be converted into orchards."[13] Chenggong County was conveniently located for the authors of the study who taught at the Southwest Associated University and resided in or near Kunming; in fact, easy access to study sites seems to have been a common feature of many of the small-scale surveys that were published in the academic journals of the time. The prevalence of such studies not only underscores the novelty and unfamiliarity of the territory to the researchers, but also demonstrates the importance they placed on collecting data about their new environments.

Perhaps because there seemed to be no overarching structure for these local surveys, the National Central University faculty authors of a different geographical survey on land utilization undertaken in the summer and fall of 1942 in Zunyi County, Guizhou, advocated for such surveys to be undertaken all over China so that there could be "a thorough understanding of the present conditions of land utilization so that measures may be taken for rational use of natural resources."[14] In other words, the authors hoped that someone would read their studies and those of others and use them as a foundation for plans on how to use the resources of the region. The article described the geography of the region, the climate, and the types of land use, making clear the percentage distribution of land in different uses (including forest, waste land, and unproductive occupation of land), types of crops, seasons of cultivation, and locations of human settlements, and commented on farming methods and overcultivation of land. In spite of the detailed description, which included material on both the common features and the idiosyncrasies of the region, the authors' aim was to produce generalizable data that might help others to better understand not just Zunyi County but also other counties with similar landscapes. As they wrote, "The present study is to set an example for subsequent research. Although the surveyed area is limited, the terrain is typical of mountainous region[s] of South China so that the condition pictured above may be taken as the general picture of land use in this part of China."[15] As in the case of the Chenggong County study, this one also finishes with recommendations about how land might be better used. Slopes, the authors concluded, were eroding too much under cultivation and should therefore be forested, and in general, "the proper and more efficient use of the land must depend on the improvement of irrigation, seeds and fertilizers, by which crop yield may be raised and winter water fields, at present lying fallow in winter, can be used for growing winter crops."[16]

The kinds of conclusions reached in such studies emphasized reconsideration of land-use patterns and implementation of new technologies, both of which had the potential to increase yield and protect the land from long-term problems such as soil erosion. The authors were universally concerned with improving the output and economic potential of the land and were often eager to propose modern scientific methods to accomplish this aim. New crop variants and fertilizers, in addition to new irrigation systems and crop rotation strategies, all had the potential to make the land more productive, but which of these would help most in any given area could only be predicted once an understanding of existing land-use patterns and agricultural norms had been established. Land-use surveys represented that first important step. The production of such surveys also served the interests of the individual authors. They were bringing their expertise to bear on an analysis of local conditions in territories with which they had previously had limited familiarity. By conducting surveys and disseminating their findings, they were both establishing their own expertise and building a body of knowledge that others could use to modernize the region.

Geological surveys of underground resources provided similar information on which bureaucrats and entrepreneurs could base decisions about where and how to develop the infrastructure necessary to access and extract those resources for industrial or other use above ground. Chinese governments did not, prior to the war, have a great track record of making data-driven decisions. For example, Grace Yen Shen has shown that geologists connected to the China Geological Survey, including Weng Wenhao, wartime head of the NRC, were already, well before the war, acutely aware of the importance of surveying and understanding the distribution of resources before producing development plans. Government bureaucrats' past failure to pay attention to geological surveys had led to catastrophic disconnects between geographical realities and government action—at least some prewar government development plans had failed to take into account the natural resource needs of the industries they were developing and demonstrated a lack of understanding of known mineral deposits.[17] Such decisions did not end with the retreat to western China. Decision makers prioritized safety from enemy attack above most other considerations when determining where to locate critical industries, with the result that numerous state-owned enterprises, including factories manufacturing weapons and other tools of war, were constructed in areas with limited access to raw materials, thus rendering them considerably less efficient than they might otherwise have been. By the early 1940s, however, government planners and technocrats were not only trying to ameliorate conditions for those plants, but they were also making a much more concerted effort to gather data that would guide their planning for future development.

As a result of this shift, both small- and large-scale surveys were primarily about the collection of data needed to make educated recommendations for technical and other improvements. Although many surveys were undertaken locally by scholars who had relocated as a result of the war, some did involve exploration of previously "unmapped" terrains. These surveys typically involved "expeditions." Expeditions involved travel, required considerably more financial outlay, were usually underwritten by the government, and were frequently undertaken by a team of scholars and bureaucrats who were working at the behest of a government organ and who sought to produce a comprehensive description and evaluation of the natural and human resources of a broad region.

Some expeditions were motivated, at least in part, by intellectual curiosity among scholars who had relocated to the west and who longed to know more about territories that had previously been difficult to access. As Grace Shen observed, "In many ways the dislocation of war was an opportunity in disguise for China's geologists. The 'mountains of Tibet' and 'caves of Kuangsi' . . . had long been frustratingly unavailable objects of fascination for coast-bound geologists."[18] The same was the case for meteorologists, botanists, other natural scientists, as well as social scientists, many of whom clearly leaped at the opportunity to explore territories about which they knew relatively little. To this end, groups like the Chinese Society of Natural Science launched their own expeditions to explore western China. In 1939, they sent a group led by geographer P. H. Chu to Xikang; in 1941, Shu-Tang Lee 李旭旦 and M. N. Jen 任美锷 led the Northwest Scientific Expedition.[19]

This second expedition aimed to explore the natural resources of the region and included Lee, a geographer, Jen, a geologist, C. S. Hao 郝景盛, a forestry expert, and S. Y. Chang 张松荫, a specialist on sheep.[20] The group departed Chengdu on July 16, 1941 and returned just over three months later, having traversed a meandering three-thousand-mile path around southern Gansu, often traveling by foot or on horseback when the roads were inadequate for their car.[21] Upon their return, Lee published a report on the expedition, and the other members published reports on the geography, geology, forests, and livestock of Gansu, all of which appeared in the January 1942 volume of *Acta Geographica Sinica* (地理學報). Lee's report included a day-by-day description of the path they took, including distances and descriptions of both topography and human geography. The other reports described the more specific features of the region, but Hao's report on the forests of southwest Gansu also offered some suggestions for developing the region.

Although Hao's report was primarily a description of the distribution of forests and types of trees in the region, he concluded it with a list of suggestions for developing the region's forest resources. In addition to ideas about what kinds of trees to cultivate and where, Hao also argued in favor of developing roads because "communications not only determine whether a development activity can be smoothly implemented, but they also determine the speed with which it can be done." In their current state, he argued, even though there were four thousand kilometers of public roads in the region, the poor quality of the roadbeds meant that "they can't bear the burden of developing the northwest." In addition to fixing and building roads, modes of transportation also needed to be improved. Horses, camels, and cow- and horse-drawn carts were the most common types of transportation, and none of them could traverse more than thirty to forty kilometers a day at best. This meant that to the extent that the region already had forest and other resources (e.g., oil and animal products), these resources could not easily be transported "outside the province or internationally," and products imported to the region "sold at ten times the price" owing to the additional costs of transportation. From Hao's perspective, then, the foundation for economic development of the region was transportation. Without improvements in that area, the northwest could not be effectively integrated into either the national or the international economy.[22]

As we have seen in the case of the Northwest Scientific Expedition, Chinese scientists fairly rapidly organized themselves for the purpose of exploring the natural environment of western China. In some cases, this meant creating new institutions that were specifically geared toward surveying the region. One such case was the China Institute of Geography (中國地理研究所), which was established as a preparatory office for an institute of geography at Academia Sinica in August 1940 in Beibei and quickly became a hub for geographic research in Sichuan and surrounding areas. Although a new branch of an existing government organization, its financing came largely from the Sino-British Institute Fund (中英慶款董事會). It was headed by the human geographer Huang Guozhang 黃國璋, who later also served on the CPB. Between its founding and 1943, the institute mounted seventeen separate geographic expeditions that generated numerous papers, many of them published in the institute's quarterly journal *Geography* 地理.[23] Other institutions spun off new branches. For example, the MEA's China Geological Survey set up a field office in Lanzhou in 1943 that it used as a base for surveying and mapping the northwestern provinces. The field office was staffed by a team of five experts on geology and mining, had a dedicated budget, and focused on the surveying and mapping of China's northwestern provinces. It also collaborated with the Gansu Petroleum Administration and related NRC entities in the region. Over the wartime years, field office personnel conducted numerous expeditions to survey the northwestern provinces.[24]

National institutions of higher education and provincial and national government entities also organized expeditions. For example, the National Central University collaborated with the Sichuan provincial government to send a scientific expedition to southwestern Sichuan from August to September 1942. The group included experts in geography, geology, botany, and agriculture, and according to an article in *Acta Brevia Sinensia*, while on their excursion, the group discovered a substantial potential hematite mine.[25] Similarly, from May to November 1943, Academia Sinica sent a group of five scholars with training in a variety of disciplines to Xinjiang. C. S. Lee, C. Lin, S. Ting, T. P. Ko, and C. Y. Chang evaluated the "geology, geomorphology, plant ecology, anthropology, ethnology and mineral resources" of the region, collected thousands of rock and fossil specimens along with numerous botanical samples and measured two thousand people. The expedition was financed by the Executive Yuan (with funding approved by Chiang Kaishek 蔣介石 himself). In addition to conducting fieldwork, the group met with local officials to discuss natural resources and sent back reports to Academia Sinica on topics such as forestry, agriculture, and mining.[26] C. S. Lee 李承三, who had in the early part of the war conducted extensive geological surveys in Yunnan at least partly under the auspices of the China Institute of Geography, collected enough material on the expedition to publish numerous articles in 1944 in several different geological journals on the geology and natural resources (including coal, gold, and other mining) of Xinjiang and the greater northwest.

Also in the fall of 1943, Franklin Ho 何廉, on behalf of a consortium of banks and the national government, took a two-and-a-half-month tour of Gansu, Shaanxi, Qinghai, Ningxia, and Xinjiang with the purpose of seeing "what the government should do and what private entrepreneurs could do in respect to the economic development of the region." He undertook the trip prior to taking up a position as deputy secretary general of the CPB (最高國防委員會中央設計局). He noted that such a trip was necessary because "except in rough outline, it wasn't known what preconditions for economic development existed there. It wasn't known what natural resources were there. We did know there were agricultural raw materials, such as cotton in Shensi and wool in Tsinghai, Ningsia, and

Kansu, for light industries, but we did not know just how adequate the supply was, its quality, or its distribution."[27] Ho's comments reveal some of the gaps in government knowledge about resources, industry, and the market in China's northwest, but they also reveal that the government was indeed gathering knowledge about western Chinese regions before trying to develop plans for their economic development. The planning-without-first-surveying approach that Weng Wenhao and members of the China Geological Survey had criticized in the early 1930s had changed.

Ho's description of packing for the trip tells us quite a bit about his expectations of the conditions he would find. "The truck was loaded with necessities—some gasoline, spare parts for the repair of the car, bedding and a cot for each person, some food, and, of course, medicine and first aid equipment. We could not be certain of getting facilities along the way, especially once we reached the northwest. We had to be self-sufficient."[28] In fact, Ho's approach to travel in the region was fairly typical. Wartime travelers to the northwest universally recounted tales of terrible transportation difficulties and trips punctuated by long breaks to wait for trucks and cars to be repaired, river portages to be arranged, and roads to be reconstructed.[29] Given the transportation problems, even for scholars and bureaucrats located in nearby Sichuan, the provinces of Gansu, Qinghai, Ningxia, and Xinjiang seemed extraordinarily remote.

Like C. S. Hao, Ho noted the poor infrastructural conditions of the region. Not only were roads inadequate, but there were also other challenges to economic development. Lack of electrical power was of particular concern to Ho, who was thinking about the region's capacity for industrial development. Without power, industry could not develop, and as a result, in Gansu, Qinghai, and Ningxia, "there was practically no modern industry to speak of." Ho also noted the comparative lack of irrigation that meant that the land could not support large-scale agricultural development. As a result, as he saw it, "the land . . . was completely barren."[30] Irrigation, power, and roads were all problems that the national government had sought to mitigate, but the area was so large and the challenges so vast that the efforts that had been made so far seemed to Ho to be little more than a drop in the bucket. Moreover, he observed that even if the central government undertook a road-building project, after its completion it was up to the provincial and local governments to maintain the road and they were not always up to the task.[31]

The purpose of Ho's expedition was to work out ways to more effectively integrate the northwest with the southwest so that the whole of western China would serve as "a second base for the development of national defense industries." To achieve this goal, CPB economist Chen Baoying argued in favor of improving the communication network that connected the two areas and setting up branches of southwestern industries in the northwest.[32] From Ho's perspective, however, simple transplantation of industry would not be a logical or productive approach to developing the northwest. In Lanzhou, he toured a woolen mill that had been built in the 1870s but that still was not producing at capacity despite having the most advanced machinery in the area because of poor transportation and inadequate supplies of both water and wool. As he later noted, "My observations at Lanchow convinced me once again that industrial development does not mean industrial transplantation. You had to begin with the prerequisites for industrial development."[33]

Based on his fall 1943 observations of China's northwestern provinces, Ho concluded that Gansu, Qinghai, and Ningxia "did not have the prerequisites for industrialization." In his view, what they needed was irrigation to develop agriculture that would provide raw materials, feed the people, and build surplus capital. To the extent that industry could be developed, the emphasis should be on light industries related to food and clothing; but to develop industry would require transportation and power. However, "no private entrepreneur could undertake to provide transportation and irrigation, the two essentials. It was up to the government. . . . The government had to make a determined effort to develop transportation, to develop irrigation, and to provide other social overhead capital for development before it even made sense to talk about the development of private industry."[34] For Ho, the expedition helped to fill in the details of an anecdotal picture that he already had of the region. Unlike other expeditions that collected and published concrete data, Ho's conclusions were more impressionistic. Nonetheless, those impressions surely shaped the way he approached the task of planning for the development of the region. At the conclusion of his northwestern tour, Ho took up his position with the CPB and put his newly acquired observations about the northwestern economy to work as he helped to develop a national program for postwar economic development.

The most extensive expedition to the northwest was the CPB's Northwest Reconstruction Investigation Team (西北建設考察團), which, headed by historian and prominent university administrator Luo Jialun 羅家倫, set off on June 7, 1943 on a nine-month, seventeen-thousand-kilometer trip through the region. The expedition was made up of a total of twenty-one members, although its long duration meant that members came and went as other things came up. Agronomist P. W. Tsou 鄒秉文, for example, was initially part of the expedition but then had to abandon it when he was sent to the United States to participate in the United Nations Conference on Food and Agriculture in Hot Springs, Virginia. Expedition participants collected materials, investigated existing development projects, and listened to reports and sought opinions from local administrators. As they got near the end of their stay in each province, the group would take time to talk through what they had seen and come up with a common set of ideas and proposals for further development. They relied heavily on local and provincial leaders as their guides and informants, and their primary goal was to look for ways to improve what was already there. The group divided its work into twelve subjects—railroads, roads, water management, agriculture, forestry, animal husbandry, land reclamation, industry, mining, health, ethnicity, and education—each of which had a small subgroup assigned to it. At the conclusion of the expedition, each subgroup drafted a report that was then edited and published by the expedition leader, Luo Jialun.[35] Unlike other expeditions that collected and published data but did not tend to go very far down the path of proposing actual plans, this one did both. The report is replete with concrete suggestions about what needed to be done and what resources would be required to do it.

As was the case with virtually every other expedition to the northwest region, the Northwest Reconstruction Investigation Team concluded that the single most important thing the government needed to do to facilitate both defense and development of the northwest would be improvement of transportation.[36] But even with good roads, new railroads, and better air transportation, development of the region, at least in the areas of agriculture, forestry, and animal husbandry, would be hampered by its arid conditions. Thus, the report argued for a logical and careful development in those fields that took into account the kinds of land and water resources that were actually available.[37] A third major issue, especially for the postwar period, would be a scarcity of trained personnel. Human resources were, in fact, an important enough concern that each chapter of the expedition's final report included a section describing the types of skilled technicians that would be required to implement development projects in that field and calculations of the total number of technicians that would be needed. Overall, the report estimated the following needs: railroad engineers: 22,300, road construction and management: 28,821, water management: 6,970, agriculturists: 3,160, forestry: 989, animal husbandry: 1,940, health: 34,088, for a total of 98,268 people of whom 20,342 needed to be graduates of colleges or vocational schools. According to the report, these people, particularly the ones with special skills, would need to be recruited to the region because the most highly skilled technicians currently in the area were mostly war migrants, and many of those people would want to return home to eastern China as soon as the war ended. To meet the region's human resources needs, Lanzhou and Xi'an would have to be turned into the university centers of the northwest, and educational opportunities in Urumqi would also need to be improved. At the same time, government organs would have to devise incentives to retain some of the war migrants after the war and to recruit new skilled personnel to the region.[38]

The report of the Northwest Reconstruction Investigation Team shows how data collection, analysis, and planning could be undertaken in concert by a well-organized and highly focused team with a clearly articulated mission. Reports produced by the other investigators we have looked at, however, demonstrate that most surveys or expeditionary studies were undertaken by people working alone or in small groups. Their findings and recommendations, though often very detailed and informative, tended to focus narrowly on a specific topic or small area. To make large-scale plans based on these sorts of studies would require collating data from multiple studies in order to get the big picture.

Such collation exercises did sometimes happen. By the end of the war, government bureaucrats, academic researchers, and developers of both public and private enterprise had collected sufficient data on Sichuan Province, for example, that a group of geographers at the China Institute of Geography were able to take Sichuan as their first case study for a provincial economic atlas, with plans to produce similar atlases for the rest of China's provinces in the future, once sufficient data had been collected. Published in the spring of 1946 but based on data collected during the war, the *Economical* [sic] *Atlas of Sichuan* contained seventy-two maps of Sichuan Province, each of which showed some feature of Sichuan's geography, climate, or economic activities (see fig. 1.1). The volume was

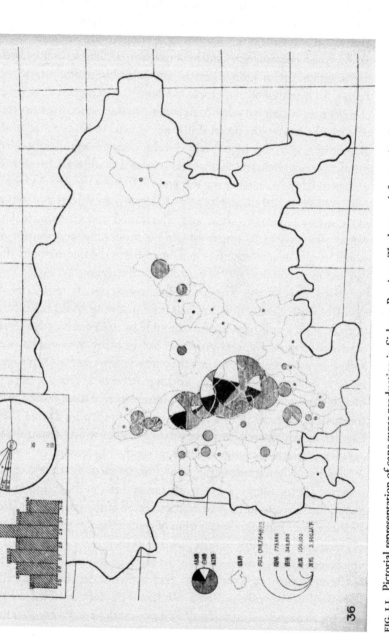

蔗糖產量分佈圖

CANE SUGAR

FIG. I.I. Pictorial representation of cane sugar production in Sichuan Province. The bottom left provides data on production in different regions of Sichuan and a key to reading the map. The inset on the top left shows annual production for the years 1936–1942 as well as production in Sichuan compared to the rest of China. Source: Zhou Lisan, Hou Xuedao, and Chen Siqiao, *Economical Atlas of Sichuan* 36

36

intended as a helpful reference for Sichuan's further economic development, though it was not produced solely for state use. For example, there were maps showing the location and density of specific agricultural products, and others showing the movement of those products into processing zones. The maps were based on more than five thousand materials collected by the authors including both published and unpublished results of research conducted by both public and private organs.[39] They were accompanied by a separate booklet that included short descriptions of the phenomena being depicted in each map as well as a series of appendices providing further information and a list of the seventy-five major sources that were used to compile the atlas.

The level of detail in this volume reveals considerable awareness of topography, landscape, climate, location of population centers, industry, underground mineral resources, land use, and agricultural features of the entire province, and the extensive list of sources that the volume relied on demonstrates just how many researchers were engaged in collecting data about the province. Although much of these data were collected by government organs, as we have seen, numerous academic teams were making investigations in both large and small localities all over western China during the war. Geographers, geologists, agricultural economists, animal husbandry experts, and ethnographers were all active in conducting surveys of land use, agricultural productivity, geological formations, industry, and human geography in the region. Their reports all served as sources for the atlas.

The atlas's pictorial representation of Sichuan's wartime economic conditions, though imperfect in that the data were sometimes for the late 1930s and at other times for the early 1940s, nonetheless effectively demonstrates the extent of efforts by bureaucrats and scholars alike to collect and catalog information about the environment in which they were living with an eye to better knowing it and also to improving it. By the end of the war, geographers had access to vastly more current information about Sichuan than any other province in China, owing in part to the relocation of both government and many institutions of higher learning to Sichuan during the war and also owing to Sichuan's extraordinary economic importance—as both breadbasket and industrial center—to the Nationalist government during the war. Over the course of the war, the government was compelled to turn to Sichuan's abundant agricultural and less abundant mineral resources to fuel a wide variety of industries and other wartime enterprises. In that context, increased production of a number of key agricultural commodities was of particular importance. As a result, production surveys, surveys that shed light on methods, practices, and markets, and development of plans to increase output were high priorities for a number of different state agencies.

Both government bureaucrats and academics, thus, spent the wartime coming to know the region to which they had relocated and seeking to understand its human and physical geography and the ways in which land and resources were utilized. This knowledge was written up and published and sometimes collated to produce a more comprehensive picture of a region than any single study could yield. As Ho's observations about his northwestern trip suggest, knowing a territory was essential to mastering it and making effective use of its resources and also to avoid making critical errors, such as attempting to turn regions with limited power supplies and poor transportation into major manufacturing hubs. Much as the expeditionary work of European scientists had supported the extension of European power and economic authority into new colonies in the late nineteenth and early twentieth centuries, so too did these wartime Chinese surveys and expeditions by scientists and social scientists provide the basis for development plans created by bureaucrats, academics, and independent technicians that almost universally sought to bring the region more squarely under state control and harness its resources for the good of the nation.

Planning

As we have seen in these descriptions of both surveys and expeditions, one of the strongest motivators for surveys was the common understanding among bureaucrats, intellectuals, and technicians of the urgent need for economic development in the region combined with a shared understanding that basic knowledge about the region was necessary to facilitate that development. Surveys provided valuable information about economic activity and the potential for improvement of agricultural output or further development of natural resources. They formed the basis for recommendations for new farming and herding methods, planting of new or different strains of crops, training of particular types of skilled workers, locations for potential industrial development, and additional research into plants, animals, and natural resources. Agricultural development of inland China was needed during the war both to provide more calories for the increased population to consume and also because agriculture formed the basis for much of western China's industrial development, supplying a wide array of raw materials necessary to fuel China's war effort. Sugar, tung oil, and oranges were all used to produce fuels, for example, whereas cotton and wool were employed in the manufacture of fabrics, including military uniforms and blankets. Similarly, surveys of both topography and underground mineral resources provided information that could be used to plan things such as the construction of new transportation networks and the development of industry. Whereas early Nationalist planning efforts had not always been grounded in the kinds of knowledge that such surveys offered, by the early 1940s, government planning was increasingly based on data that had been gathered by both government and nongovernment groups.

GOVERNMENT PLANNING

Otis Graham, writing of planning in the mid-twentieth-century United States, has observed that "planning assumes that modern industrial society requires public intervention to achieve national goals; assumes that such intervention must touch all fundamental social developments; must be goal-oriented, and effectively coordinated at the center; must be anticipatory rather than characterized by ad hoc solutions and timing dictated by crisis."[40] In spite of the fact that it was operating in a crisis situation, these same traits characterized the planning ethos of the Nationalist government during the war. By 1940, the Nationalist government had identified planning as the first of three pillars of its administrative approach. The triple system of administration, promoted throughout the bureaucracy by Chiang Kaishek, aimed to improve efficiency and extend state control over more resources.[41] As Chiang observed that year in a speech to government officials, "No matter what kind of work we are doing, we must plan, execute, and assess, and we must do these things coherently and not disjointedly, only then can we carry that work through to a perfect end. . . . Any government organ must undertake its work in accordance with the three pillars of planning, execution, and assessment in order for its industrial efforts to continuously improve and produce real results."[42]

As a consequence of this approach, the Nationalist government developed what, on paper, looked to be a fairly streamlined process for planning the government's yearly activities. In October 1940, it created a CPB (中央設計局) that was to be responsible for both formulating and studying all political and economic reconstruction plans.[43] At the beginning of each year, the National Defense Council (最高國防委員會) would issue a set of general goals for the year that would be sent on to all government ministries. Within each ministry's Committee of Planning and Examination, plans to meet those goals would be drawn up and forwarded to the CPB. The CPB would then organize all of the ministerial-level plans into a national plan for the year; however, in some years, such as 1942, all it could really manage to do was to prioritize the ministerial-level plans so that funding would be channeled to the highest priorities. Starting in 1943, in addition to annual planning, each ministry was required to draw up a ten-year plan, and doing it well necessitated considerable coordination with all related government units. These plans, too, were to be submitted to the CPB. Like the yearly planning system, this strategy was also hard to fully implement. By August 1943, for example, some ministries had reportedly completed their plans, but others had not yet written them.[44]

In addition to the plans mentioned above, the CPB also undertook a series of special planning projects. For example, in 1941, it compiled a three-year wartime reconstruction plan; in 1943–1944, it began working on a ten-year plan to develop China's northwest, and it produced an outline for a demobilization plan in the spring of 1944.[45] In 1941, Chiang Kaishek asked the CPB to draw up a postwar five-year plan for national defense and economic reconstruction,[46] though it appears that work did not begin in earnest on this plan until 1943. According to Franklin Ho, who was appointed deputy secretary general of the CPB in 1944, it did not really get going until after he joined the group. In other words, the CPB was expected to develop a wide array of both short- and long-term plans, but there is ample evidence to suggest that its capacity to fulfill those expectations was limited, at least in the institution's early years.

Franklin Ho was brought into the CPB in 1944 as "the technical man in economic planning."[47] In that capacity, he reorganized the CPB into departments, each of which focused on a different economic sector—communications, power, mining and metallurgy, agriculture, water conservancy, and manufacturing industries—and appointed people in leadership positions from the various government ministries and bureaus connected to those sectors as chairs of the new departments. Through this method, according to Ho, he positioned the CPB to better access the cooperation it would need from the ministries and bureaus and, as a result, to better accomplish the planning tasks it was assigned to undertake.[48]

Guiding the long-term planning work of the CPB was a committee that Ho put together to determine "principles of economic enterprise" that included geologist Weng Wenhao 翁文灝, who was also minister of economic affairs and head of the NRC, Minister of Communications Zhang Jia'ao, historian and diplomat Jiang Tingfu 蔣廷黻 who was at that time the political director of the Executive Yuan, Lu Zuofu 盧作孚, a Sichuanese entrepreneur who was also vice-minister of communications, H. H. Kung 孔祥熙, minister of finance and governor of the Central Bank, and Hsu Kan 徐堪, chair of the KMT's finance committee. The major focus of debate in this group was the question of the extent to which industry should be under the control of the government.[49] In the end, the committee advocated for "planned development in a mixed economy."

Once the basic principles had been agreed on, the five-year postwar reconstruction plan itself was fleshed out by a group of seventy bureaucrats who worked in subcommittees on specific topics (e.g., mining and metallurgy, communications, agriculture) over the course of 1944. They gathered data and five-year plans from the relevant government organs, compiled this information in numerous tables, and used it to project future development goals in the areas of communications, mining and metallurgy, industry, agriculture, and water management, along with the financial and personnel requirements necessary to meet those goals. The resulting two-part plan aimed to establish a basic foundation for industrialization, meet the fundamental requirements for national security, and raise the standards of both health and knowledge of the Chinese people. Once efforts to achieve those goals were under way, an additional set of reforms to economic and political structures would be undertaken. The plan placed equal emphasis on defense and general economic development and divided the nation into nine economic regions, each of which was to be developed based on its current conditions so that, for example, the central region would have the most industrial development, whereas irrigation and hydrology would be the main focus in the northwest. The authors of the draft plan intended that it be implemented starting on July 1, 1946, but as it turned out, the year spent in its development was largely to naught, as the plan was never carried through and the CPB was disbanded.[50] Regardless of whether the plan was actually implemented, the process through which it was produced, in combination with the other activities of the CPB, demonstrates both the extent to which the Nationalist government was permeated by a planning ethos and the degree to which it perfected its planning infrastructure over the course of the war.

Important to note is that even as the postwar reconstruction plan was being produced, there was not universal agreement within the government over exactly how rigidly the government should follow the plan once it began implementing it. According to John D. Sumner, economic analyst with the U.S. embassy in Chongqing, Franklin Ho and political scientist and KMT politician Gan Naiguang 甘乃光 were thinking of the plan as a detailed blueprint for economic development that would include production targets, location and financing of plants, and even details on the particular types and quantities of products each plant would manufacture. Weng Wenhao, on the other hand, still concerned even in 1945 about the national government's emphasis on planning, saw this level of detail as "impractical" and viewed the plan as more of a general guide. In other words, even though government planning was, by 1945, much more solidly grounded in data than it had been a decade or more earlier, government planners may still have been putting more faith in their plans and in the seemingly unrealistic possibility of executing them exactly as written than was merited.[51]

The bureaucratic inclination to produce plans was strong enough that individual bureaucrats who were sent abroad to study also sometimes developed plans for the improvement of the specific industries within which they worked while they were abroad. One example can be seen in the work of Peng Wen-ho, a senior technician in the Chinese National Research Bureau of Animal Industry who was sent for training in June 1945 to Iowa State College, University of Wyoming, Colorado State College, and Oregon State College. Peng's area of specialization was sheep and wool, and he had managed the Southwestern Sheep Experiment Station in Yunnan prior to his training period in the United States. Upon completion of his one-year training program, Peng submitted a report in which he outlined a proposed program for sheep improvement in China that was based on conclusions he had drawn from his coursework and site visits in the United States along with his personal experience with the subject in China.

The objectives of Peng's plan were to "increase the national financial income and coordinate the national industrial wool demands" by increasing production of higher value clothing wool as opposed to the more commonly produced and lower value carpet wool; increase income for sheep raisers; and introduce new sheep breeds to different regions that would be best suited for the environments of those regions. To achieve these objectives, Peng proposed introducing specific foreign breeds of sheep to different regions where they would improve existing sheep stock through interbreeding and artificial insemination and where pure breeding would need to be undertaken on a large scale by regional sheep experiment stations under the National Wool Administration of the Ministry of Agriculture and Forestry (MOAF). He further proposed that these experiment stations improve range management by, among other things, developing better watering systems, and encouraging herders to rotate grazing and defer grazing until later in the spring. In addition, he proposed that experiment stations work with herders to develop better winter feed provisions so as to limit loss of livestock due to starvation in the winter. Peng also advocated for the creation of a national wool marketing corporation and cooperative wool growers' associations that could work together to improve wool handling and marketing. Finally, he proposed that research on standardization and grading of wool be undertaken by both national and local wool laboratories.[52]

Peng's proposal, though narrowly focused on a specific agricultural industry, was targeted at extending state control over that industry as well as improving the national economy and the livelihood of farmers engaged in it. It did so by mapping out a comprehensive set of activities that, taken together, had the potential to transform and dramatically improve China's wool industry. In Peng's view, it was the state's duty to implement the needed improvements, and he imagined a new system of state organs that would carry out those improvements. He planned for at least four sheep experiment stations, each in a different region of China (there were at the time only two: one in Yunnan and the other in Gansu), an extension service to train technicians and establish wool marketing associations, a national sheep breeders' association that would help establish local breeders' organizations, and a wool laboratory. All of these institutions would fall under the umbrella of a national wool administration, housed in the MOAF.

Peng had been sent to the United States by the government to study both wool and dairy and "to prepare himself to better serve China." As a government employee who had himself worked for a state agency and who had been sent abroad for further training, it is not unusual that he should have produced a plan with the specific intention of submitting it "for his government's consideration and possible adoption";[53] in fact, it is quite likely that his employers expected him to generate some kind of plan or at least to be thinking about how he might apply his training upon his return to China. It appears from his wording, however, that Peng may have gone above and beyond those expectations and had in fact taken it upon himself to develop his plan as a special project at the University of Wyoming, where he had found a suitable mentor in Dr. R. H. Burns, himself a wool researcher.

Wool, as a major prewar export commodity and one that was garnering high prices from the Japanese in north China, was of some concern to the government, which had in May 1943 issued a set of regulations governing its purchase and distribution. These regulations, which specifically targeted wool smuggling, mandated that wool merchants in China's western and northwestern provinces register with the government. They also required that all wool exports be run through the government-controlled Fuxing Trading Corporation. They did not, however, address the issue of wool production.[54] Researchers in Lanzhou, including the Americans Ralph W. Phillips and Ray G. Johnson, who spent a good deal of 1943 in the region working with livestock, undertook studies on sheep-raising practices and reached many of the same conclusions about pasturage, winter feed, and water as Peng.[55] However, there is no indication that anyone in China was developing a national wool plan. To the contrary, according to Phillips and Johnson, based on their own observations while working in the field for a year as well as data supplied to them by Chinese government agencies, China suffered from a scarcity of trained experts in animal husbandry and a comparative absence of information about livestock numbers and conditions. Phillips and Johnson also concurred with Peng's conclusion that the quality of Chinese wool needed to be improved.[56]

Peng's plan, along with others that can be found in the files of Chinese government technicians sent to the United States for technical training at the end of the war, demonstrates that the planning ethos had permeated the bureaucracy to the level of the individual.[57] Not only did bureaucrats generate plans at the command of their managers, ministers, or the CPB, but they also came up with plans when they were off on their own, in other countries, honing their technical skills and acquiring knowledge with the expectation of putting their training to work for the nation. It should come as no surprise that having come from a bureaucratic environment that emphasized planning, individual technicians sometimes felt that the most appropriate and meaningful way to apply their learning (particularly from afar) was through articulation of large-scale plans for their particular industries.

PLANNING IN THE ACADEMIC REALM

Peng was not the only planner to be thinking about institutional structure. This was also a common feature of agricultural plans during the wartime period. For most of the agricultural experts who undertook to construct plans for agricultural development, knowledge of agricultural conditions was just one part of the planning puzzle. The other parts, perhaps even more important, were institutional structure and agricultural education. Although some agricultural experts spent the war years collecting data on specific dimensions of plant pathology, planting strategies, irrigation, and land use, other agricultural experts thought less about the farming itself and more about the structure of government organizations and human resources that would best support farmers and guide them toward best practices that were based on the findings of surveys of the sort mentioned above.

In 1943, P. W. Tsou, president of the Agricultural Association of China and former dean of the College of Agriculture of National Southeastern University, and C. W. Chang 張之汶, who was then dean of agriculture of Jinling University, looking to ensure that agriculture would not be forgotten in government planning for the postwar period, produced a Program for Postwar Agricultural Reconstruction in China. The December 1943 proposal represented the views of the membership of the association, having been edited in response to critiques members had made in meetings in

Chengdu and Chongqing and through correspondence for members who were unable to attend. In April 1944, Tsou, who was by then in the United States, translated it into English and made, on his own, some additional edits before sending it to the printer to produce enough copies that he could disseminate it to both Chinese and American agricultural specialists in the United States in hopes of getting further comments. Fearful that agriculture would be left behind in the national planning process, Tsou and Chang argued a point that Tsou made frequently in speeches and essays: that the two were so intimately intertwined that agriculture simply could not be ignored in any race to modernize China. "After the war," they wrote, "China will launch a large-scale reconstruction program to build the country into a modern nation as other powers of the world. Her national resources must be developed to enrich the country and to elevate the peoples' living. A question was raised as to whether industry or agriculture should receive major emphasis in the postwar reconstruction. This is a point beyond debate, since one needs to be developed as much as the other. Furthermore, they are interrelated and mutually dependent." Their program was designed with an eye to developing agriculture in ways that would facilitate and support China's industrial modernization.[58]

To accomplish this goal, the association proposed that China's national agricultural policy should aim to provide for the people by increasing food production, producing a government plan for land utilization, protecting farmers' rights and interests, and improving nutrition. At the same time, it should use agriculture to help develop industry by promoting large-scale production of agricultural products with industrial uses, improvement and expansion of agricultural exports, and expansion of farming in border areas.[59] To accomplish these goals, they proposed that the government should undertake comprehensive surveying and registration of land, require land owners to consolidate land holdings so as to have larger-size farms, rather than having scattered small plots that were agriculturally inefficient, mandate land use depending on its topography, nationalize spent land to take it out of use for the purpose of conserving water and soil, and fix farm tenure problems by bringing legal order to relations between landlords and tenants. In addition, they called on the government to advise and direct farmers on what to plant, how to improve both quality and quantity of production, how to promote the improvement of farm implements, organization of labor, and structure of farms, and to promote the organization of rural industrial cooperatives to utilize excess product and employ farmers in slack seasons. They also called on the government to improve transportation networks to facilitate both domestic and international trade, develop a method to standardize and grade agricultural products, and create an agricultural inspection system.[60]

Agricultural policy was, in their view, the domain of the government. The government, they argued, should take responsibility for bringing order to agriculture by conducting surveys, implementing laws and regulations, promoting certain practices through education and mandate, and establishing infrastructure. Through these mechanisms, agricultural output would increase and the standard of living of farmers as well as the overall economic situation would improve. To round out their plan, the association laid out an institutional structure that they believed would be best able to implement these reforms. They called for "a linear system of coordination" between the organs, "a clear division of responsibilities," "strengthening the functions and powers" of all of the organs, and "close coordination and cooperation" between the organizations.[61]

Even actors operating outside of the government bureaucracy, therefore, sought to contribute to the national good through a process of planning. But as they did so, they made clear their expectation that while government outsiders could and should contribute to the planning process, the execution of development plans should be the domain of the government. To that end, plans almost always appear to have involved, as did Tsou and Chang's, suggestions for reorganization of institutional structures that currently acted as impediments to development and modernization. Government had the potential to facilitate and oversee the kind of development China needed, but to do so would require reform and implementation of the right kind of plan.

UNSOLICITED PLANS

Small groups and sometimes even individuals also sought to participate in the planning process, at times because they were convinced that the government was not actually conducting activities according to a coherent or logical plan. One such individual produced a plan to improve alcohol production that he sent to Eugene M. Stallings, a liquid fuel expert sent to

China in 1945 in conjunction with the American War Production Mission in China. The author, Shang-cheng Pan 潘尚貞, was a University of Wisconsin PhD in biochemistry who had most recently been working as chief engineer at the Lengshuitan Distillery in Hunan, but who in early 1945 was living as a refugee in Pingyu, Guizhou. Having completed his doctorate in 1939, he had returned to wartime China where he spent 1940–1943 in Yunnan teaching at Southwest, doing research at Tsinghua, and serving as an advising engineer at the Heng-Tung Distillery in Kunming.

During this time, Pan had observed that although the alcohol industry was critical to the production of fuel in petroleum-poor western China, and in spite of the fact that to meet these needs, the alcohol industry "has been extensively developed in interior China . . . neither is it developed in accordance with a kind of 'national fuel program' nor is it working on a well founded industrial basis." His aim was to address this planning deficit. Pan's thirty-six-page plan, copied out carefully in neat English writing, aimed at "solving certain technical problems involved in the alcohol industry so as to render the industry developing on the principle of utilizing the nation's natural resources for her war needs."[62] His suggestions were based on his own laboratory experiments and the observations he made while working in Chinese distilleries and visiting American ones before his return to China. In addition to recommending specific fermentation methods based on his research, Pan also included in his plan suggestions for the number and type of distilleries and the sorts of equipment to be used and argued also that experimental plants, where students could be taught to establish and manage distilleries, should be set up in major population centers.[63] Perhaps Pan's plan, written while he was between jobs, probably desperate for employment and hoping to be soon notified of a new placement that he expected to get at the NBIR (經濟部中央工業試驗所) in Beibei, was little more than a self-serving effort to find a position with the War Production Board. After all, he used his concluding paragraph to offer his own services to conduct the necessary experimentation to determine if the kind of experimental plant he proposed would actually yield the results he anticipated.[64] However, the thoroughness of his proposal also demonstrates that this was a question he had considered extensively and that he was eager to get his ideas for improvement of the alcohol industry to people who might actually have enough influence to make a difference. More importantly, Pan and his plan show us just how deeply the impulse to plan had permeated the way of thinking of Chinese technicians. For Pan, who had recent experience working in state-owned industry and academia but who was now living as a refugee and trying to find a way to better his own life and participate in the life of the nation, planning offered a path forward. It enabled him to put his expertise to some use and to find a way to participate in the life of the nation as it fought for survival.

PLANNING OUTSIDE OF CHINA

Planning for increased production, more streamlined production processes, more coherent and useful administrative structures, and the training of more personnel to undertake these activities was a signature feature of Nationalist-controlled wartime China, but it was an activity that many overseas Chinese also participated in. For Chinese abroad, planning was one way in which they could contribute to national reconstruction and resistance against Japan. There were also occasions upon which the Nationalist government sought to encourage this planning impulse among educated Chinese abroad, presumably with an eye to both harnessing their brainpower and encouraging their allegiance. One such case occurred in 1942, when Chih Meng 孟治, director of New York's China Institute and a nonaligned intellectual living in the United States, wishing to secure from the Chinese government a fund to support Chinese students stuck in the United States during the war, approached Finance Minister T. V. Soong 宋子文 with the idea of expanding a scholarship program that Meng had already begun to develop with U.S. government assistance. After some deliberation and delay, caused largely by H. H. Kung's suspicions that Meng was not a politically appropriate partner for the enterprise, Soong and Kung finally agreed to provide the support on the condition that it be granted within the context of a new group—the Committee on Wartime Planning for Chinese Students in the United States (留美中國 學生戰時學術計畫委員會).

The new committee had, as Meng wrote in his autobiography, "five honorary chairmen, two honorary vice-chairmen, the usual array of officers, and twelve members—in sum, an imaginative mix and subtle balance of Chinese interests and influences at home and abroad." Soong, Kung, former ambassador to the United States Hu Shi 胡適, Minister of

Education Chen Lifu 陳立夫, and current ambassador to the United States Wei Taoming 魏道明, all of whom held government posts and most of whom were KMT ideologues, were the honorary chairmen. The officers and members, however, also included a number of bankers, industrialists, agricultural experts, legal experts, diplomats, and academics, many of whom had ties to the KMT but not all of whom could be described as ideologues. As Meng put it, "There were important men to lend weight to our endeavors, Kuomintang errand-runners to see that we did not go astray, bankers and businessmen to check our unworldly, academic impulsiveness, and just enough serious workers in our specialized field to get the job done."[65] Meng himself was named executive director of the new committee.

Its very name—Committee on Wartime Planning for Chinese Students in the United States—indicated that the group had not been designed to simply administer a scholarship fund. The proposition that it was appropriate and necessary to plan was posited in that name, one that had been devised by Nationalist government leaders, and that proposition was accepted by the committee itself as well as the numerous members of the national reconstruction forums that the committee helped organize on college campuses across the United States in 1942 and 1943. These forums met regularly to discuss matters related to China's reconstruction. One focus of these discussions, as we shall see in chapter 3, was human resources training and development. However, an equally important concern, one that encompassed the question of human resource development, was that of how best to plan for China's future. The midwestern conference of the National Reconstruction Forum held in Saint Paul in August 1943, for example, made three resolutions related to planning: "(1) The conference suggests a balanced program of reconstruction in China after the war, including equal emphasis on industrialization, agriculture, commerce and mining, with modern techniques adopted as far as possible. (2) The conference advocates an increase in the effort for the training of personnel in all fields of post-war reconstruction. (3) The conference urges that Chinese students prepare themselves not only in the specific fields they are interested in, but also to acquire a broad outlook on all fields and to acquire a pioneering and enterprising spirit."[66] That there should be a "program of reconstruction" for postwar China was taken as a given. The only question was what the dimensions of that program should be.

A look at articles in the Committee on Wartime Planning's *National Reconstruction Journal* that grappled with planning shows the diverse range of perspectives on specific elements of postwar reconstruction plans and, at the same time, a general agreement that planning was needed. Not all participants in these discussions were in complete agreement about what features a comprehensive economic development plan should have. As was the case in the CPB, nationalization of industry, for example, was one topic on which there was no universal agreement even among people who were generally supportive of the national government. The diversity of opinion on this topic is evident in the pages of the *National Reconstruction Journal* where T. P. Hou 侯德榜, himself a captain of the chemical industry, argued against it, saying that "government alone cannot develop the national wealth in a country." Li Ming, a contributor to the journal, advocated in favor of soliciting foreign investment to facilitate the development of industry. Engineer L. F. Chen 陳良輔, who was employed by the NRC's New York office, argued in favor of nationalization, though he proposed that government should focus first on developing heavy defense industries, as opposed to lighter civilian industries, and suggested that although light industry should remain in the private sector, it should nonetheless be subject to government regulation.[67] These men all traveled in the same intellectual circles in the United States, participating in the activities of the CIE and contributing to the *National Reconstruction Journal*, but they did not agree on everything and took advantage of the opportunities afforded by CIE meetings and publication of the journal to express their different views in hopes that they would be heard in Chongqing, though perhaps also with the expectation that they would be heard in the United States. By 1944, as we have seen, the Nationalist government had made clear its commitment to nationalizing heavy industry and had also delineated a set of regulations (but no restrictions) for the utilization of foreign capital (another subject that was debated in the pages of the *National Reconstruction Journal*).[68]

Also assumed in both planning proposals and debates over planning that were published in the *National Reconstruction Journal* was that the Nationalist government would be playing an important role in postwar reconstruction. A statement made by the CIE following its spring 1943 forum on postwar industrialization of China, observed that "while other nations will be primarily concerned over the restoration of their pre-war industries, China's task will be immeasurably more burdensome, for besides rehabilitation work, she must further seek to raise herself far above the industrially backward position she occupied before the outbreak of the war. Such a gigantic undertaking requires careful and comprehensive planning."[69] L. F. Chen, who was president of the CIE in addition to his position as NRC representative in New York during the war and who also participated in the national reconstruction forums, elaborated on this point and advocated for what he called "close and congenial cooperation between the government and the people" so that "China's industrialization [can] be accomplished with the quickest and maximum result." This cooperation should, according to Chen, come in the form of a national plan. He observed that

China's post-war industrialization will unquestionably require the same kind of centralized planning as Russia did. The immensity and complexity of such a national plan as well as the thoroughness and accuracy it calls for can well be imagined. It will require the pooling together of the best brains from all professions and all ranks to study, design, and formulate the thousands of details before a well balanced and really workable plan can finally be evolved, ready for execution. Such centralized planning can only be done by the government with the whole-hearted cooperation of all the people. While no effort should be spared to tap the knowledge, experience and ingenuity of the specialists and experts along all lines in and outside the government as well as in and outside the country, the centralized supervision of the planning work by the highest governmental authority as vested in a National Supreme Planning Council or Board will be absolutely necessary.

The process of developing a national plan, according to Chen, should be inclusive and thorough, and should involve consultation with experts in a wide range of fields. "If necessary," Chen went on to say, "services of foreign engineers and consultants may be employed to assist in the planning." But it also needed to be overseen by a government group constituted by "a competent staff of specialists, engineers, and assistants, all devoting full time and attention to planning work."[70] Chen's words echoed both the planning ethos of the Nationalist government and the thinking that pervaded Weng Wenhao's MEA and the NRC. This is not surprising given that Weng was also president of the Chinese Institute of Engineers and Chen was an NRC employee. Chen envisioned a planning process coordinated by a highly competent and well-staffed state-run planning bureau that would gather advice from government, non-government, and foreign sources, and that would make sense of these many pieces of information and advice to produce an effective plan. Ultimately, in his view, that plan should be not only produced but also administered by the state. Even so, Chen imagined a process that was broadly inclusive and even transnational.

This notion of inclusivity and transnationalism was echoed in other articles in the *National Reconstruction Journal* and also characterized the process through which actual plans were constructed by groups located in China such as the Chinese Economic Reconstruction Society and the Agricultural Association of China.[71] The national reconstruction forums, which were, after all, held in the United States, explicitly invited both American participants and Chinese residing in or visiting the United States to propose their own ideas about how China should approach postwar planning. Contributors to the journal, who were also living in the United States, were well aware of the role that state-led planning had recently played in the development of the Tennessee Valley Authority, for example, and published reports on the results of U.S. government planning efforts that they observed.[72] The Chinese Economic Reconstruction Society and the Agricultural Association of China, which unlike the Committee on Wartime Planning, actually did produce concrete plans that they hoped the government would adopt, both involved large groups of contributors, most but not all of whom were Chinese. Both groups also made their plans available in English as well as Chinese, perhaps, in the case of the Chinese Economic Reconstruction Society, merely to further publicize their ideas, but in the case of the Agricultural Association of China, the intent of publishing in English was also specifically articulated as enabling the association to present the plan to "American friends and fellow Chinese agricultural students in the United States for criticism and suggestions."[73] Even in the case of individual and independent planners like Peng and Pan, we see evidence that they were mobilizing their own international training and experiences in the service of China and that Pan, especially, was also trying to engage with a transnational network of expertise by writing his plan in English and sending it to the American adviser Stallings.

Conclusion

The kinds of surveys, expeditions, and plans described in this chapter aimed to address both immediate and long-term needs. Through accumulation of knowledge and development of data-driven plans, both individual scientists, technicians, and bureaucrats, and the Nationalist government hoped to solve wartime problems and set China on a path to becoming a modern industrial society with an integrated national economy. However, these processes also aimed to bring these regions under more direct state control. Many of the plans that were developed during the war were not executed, but all of them involved the creation of institutions to guide and then regulate specific economic sectors.

During the war, individual technicians and technically oriented groups and societies in China and the United States bought into the planning ethos that permeated the Nationalist government and looked for ways to bring their expertise to bear on solving specific technical problems that were of concern (or that they felt should be of concern) to that government. This way of thinking, as we have seen in the case of the Committee on Wartime Planning, was encouraged by leading members of the Nationalist government. For many technical experts in China as well as Chinese students and overseas Chinese in the United States, planning was a way to utilize their expertise, give shape to their research activities, and enable them to contribute (or at least feel like they were contributing) to the betterment of what surely must have seemed like a somewhat hopeless and chaotic situation. Either stuck in the United States without being able to provide concrete assistance to the war effort or having retreated to regions where both transportation and industrial infrastructure was inadequate to support a modern economy and where shortages of basic goods were not uncommon, these technicians were primed to be thinking about how to make things better. Some, as we saw in the cases of Peng and Pan, focused their attention on very specific problems about which they had developed some expertise. Others, like some of the contributors to the *National Reconstruction Journal* and Tsou and Chang of the Agricultural Association of China, were concerned with larger systems and structures. Although all of the Nationalist government's own organs engaged in planning exercises, and it created in 1940 a central organ, the CPB, to oversee the construction of comprehensive plans, there is no question that the Nationalist government was also actively encouraging nongovernment experts to engage in planning as well. Through entities like the Chinese Institute of Engineers and the Committee on Wartime Planning, government leaders pulled nongovernment experts into planning discussions. Similarly, the Nationalist government also encouraged participants in government-funded exploratory groups such as Academia Sinica's Xinjiang expedition and Luo Jialun's northwest expedition to plan for both utilization of resources and more effective integration of China's border provinces into the national economy.

One thing that many of China's surveyors and planners, both government and nongovernment, were concerned with was just who would implement all of the activities they were planning. This key theme can be seen in many of the wartime development plans that came from outside of government, but it was a point of emphasis among government actors as well. As mentioned earlier, every chapter of the extensive report written up by Luo Jialun and his team on their northwest expedition included a section on personnel needs. Similarly, Tsou and Chang's plan for agricultural development also focused on the importance of developing appropriate human resources. The plans created by the CPB did the same. Human resource development was a critical piece of the reconstruction puzzle and one that occupied a great deal of attention from the Nationalist government and its allies abroad. The next two chapters will take up the question of human resource development and will focus, in particular, on transnational strategies to train the nonmilitary technical personnel that China needed both in wartime and in preparation for postwar reconstruction.

CHAPTER 2

Developing Human Resources

Education and Training in China

In his oral history, Li Hanhun 李漢魂 complained that one of the major obstacles he faced as governor of Guangdong during the war was his inability to get good technical personnel. "The lack of special or technical personnel to meet wartime demand was . . . [a] problem. Time did not permit us to cultivate enough personnel to fulfill the need. Nor did the provincial government have the financial capacity to carry out any large-scoped training program."[1] Governor Li was not the only one to make this observation. In fact, a critical feature of nearly all development proposals that emerged in inland China during the war was human resources development. China lacked sufficient trained personnel either to implement government projects in the short run or realize projects that planners were developing for the future. Therefore, a central focus for wartime planners was development of the personnel needed to meet these demands. The Ministry of Education and the National Resources Commission (NRC) were particularly engaged in this endeavor as they attempted to revamp university-level education, expand opportunities for study abroad, and set up training programs both in China and abroad. However, other government and nongovernment groups also established an array of technically oriented domestic training programs for Chinese during the war.

By and large, these educational and training strategies aimed at increasing the supply of technically skilled personnel that could facilitate specific development goals in industry and agriculture. For some nongovernment groups, those goals were very concrete, local, and specific. For the government, however, the aim was to train large numbers of people in fields that had clear and immediate application in western China and that would also be useful to the reconstruction process that Chinese officials imagined would take place at the end of the war. To meet the pressing needs of wartime and to prepare for a postwar world, the Ministry of Education attempted to reshape the system of higher education while the NRC, some military units, and other government agencies such as the Ministry of Agriculture and Forestry (MOAF) attempted to create training programs to develop skilled technicians who could assist in specific development projects. In addition, nongovernment actors such as the China Industrial Cooperatives (CIC) also sought to train skilled laborers who could contribute to economic development on a very local level. Finally, the government negotiated with foreign governments, in particular the United States and Great Britain, to bring technical experts to China to provide both training and advice.

Not everyone agreed that a pragmatic restructuring of the educational system that oriented it toward applied rather than theoretical studies was the best approach. In fact, not even everyone working for the Nationalist government was convinced of this. Hu Shi, Chinese ambassador to the United States in the early years of the war and a humanist, expressed to staff of the U.S. State Department's Division of Cultural Relations that because of the "emphasis on the technical . . . many branches of liberal learning were being neglected."[2] Yueh-lin Chin 金岳霖, a logician from the Department of Philosophy at Lianda who was invited to the United States by the Department of State in 1943, made it clear in a meeting held at the University of Chicago that he disapproved of the trend toward applied sciences at the expense of basic science. The problem, he noted, was that students in applied sciences such as engineering were not being adequately educated in foundational subjects such as physics. "I think the importance of the allied subjects is not sufficiently emphasized. Consequently, I feel that, even if we establish a research bureau for aeronautical engineering, we shan't succeed because there is not sufficient background and training, not sufficient attention paid to the allied subjects. You see, all these things are new to China—physics, chemistry—all those things. We haven't a tradition of physics and chemistry and the pure sciences. We have to develop and encourage those in order to be able to bring those things up to a certain standard, and at the present time it seems the danger is that before we attain to any kind of standard along those lines, we shall be so discouraged—these things will be so discouraged—that very few people will go in for them. I think if we overemphasize engineering as engineering and underemphasize physics and chemistry and the pure sciences, we shall not succeed."[3] Chin felt that without a foundation in the basics, students would not be able to meet more complex challenges.

Chin's travel companion, Chiao Tsai 蔡翹, a physiologist from National Central University, saw things differently, observing that "I think there is a tendency in the initial period of industrialization to copy what has been done in other, foreign, countries. So I think that the importance of physics and chemistry will naturally follow. The same situation will occur in medical schools. I am not at all pessimistic. . . . A group of pure physicists will never get into engineering. You have to start engineering, and then create the physics later. The products of the factory will have to support the research in physics and chemistry."[4] In his remarks, Tsai echoed the dominant thinking of his time, that China needed to develop quickly and in response to immediate needs. Such development could not always be grounded in a firm and thorough understanding of the fundamentals, but with time, following upon success in the initial stages of development, the nation would have the luxury to expand into the basic sciences.

Human Resources Needs Projections

As we have already seen in chapter 1, government bureaucrats tasked with planning for both immediate and future development were very concerned with the question of technical personnel. This concern with human resources was nothing new for the Nationalist government. Chinese government documents from the early to mid-1930s are replete with references to the need to train various kinds of personnel in fields ranging from aeronautics to telegraphy to drinking water to finance.[5] This way of thinking about human resources planning continued into the wartime period with regard to both short- and long-term needs, and planners thinking about postwar economic development were particularly concerned with making sure China had the technical personnel it would need to begin modernizing the Chinese economy as soon as the war ended. Both smaller, regionally focused development plans such as that produced by Luo Jialun's Northwest Reconstruction Investigation Team and national development plans such as the Central Planning Bureau's (CPB's) five-year plan included extensive and detailed projections for human resources needs.[6]

The CPB's First Five-Year Economic Development Program anticipated a postwar need for administrative personnel with training in the fields of business administration, personnel management, finance and accounting, and technical personnel in a wide range of engineering fields as well as physics, agriculture, and forestry. The plan called for the construction across China of a mixed economy within which the state would be responsible for infrastructure, certain types of large-scale state-run industries, public utilities and water conservation, and agricultural improvement, experimentation, and extension. It calculated the technical and administrative personnel that would be required during the five-year period to construct and run these entities, grouped them into three grades on the basis of level of education plus on-the-job experience, and divided less-educated laborers into the categories of skilled and unskilled. Altogether, the plan called for a total workforce of 4,968,093 that would be constituted of 4.4 percent administrators, 5.07 percent technicians, 23.87 percent skilled, and 66.66 percent unskilled laborers. Of these, nearly 60 percent would be employed in the broad field of communications and transportation, 1 percent would work in the power industry, 10.2 percent in mining and metallurgy, 18 percent in manufacturing, less than 1 percent in agriculture, and 10.3 percent in water conservancy. Even if we assume that unskilled laborers required no special education or training to do their jobs, the plan still called for 470,437 administrators and technicians, all of whom would be required to have at least a high school education, and over 40,000 of whom would be expected to have either a college education or a minimum of eleven years of on-the-job experience in addition to a high school degree. The plan also called for 1,186,222 skilled laborers for a total of 1,656,659 technicians, administrators, and skilled laborers distributed across China's various economic districts in numbers that would serve the specific needs of those districts.[7] These numbers, especially of technicians and administrators, seem higher than what the nation actually had, and even if there were that many technicians and administrators, they were not distributed equitably across China's economic regions, nor were they all trained in precisely the fields that the plan called for. The human resources needs of reconstruction were going to be large and those personnel would need relevant training and could not just be concentrated in coastal regions.

As mentioned in chapter 1, regionally focused surveys such as that undertaken by the Northwest Reconstruction Investigation Team proposed specific steps for developing the regional educational infrastructure necessary to produce the requisite highly skilled personnel to modernize the region. However, planning for such human resources development in China's northwest was only just beginning in 1943 and 1944. For example, Luo Jialun's 1944 *Report on the Northwest Reconstruction Investigative Expedition* included lists of how many skilled and unskilled workers as well as technical experts would be needed in the northwest to address each of a series of fields such as road construction, forestry, and water management. In public health, for example, the team calculated that for every nine thousand inhabitants of the region, there should be at least one doctor, one midwife, three nurses, and six additional medical personnel. In addition, sanitation engineers, pharmacists, and laboratory technicians would be needed. In total, the group proposed that the northwest (Shaanxi, Gansu, Ningxia, Qinghai, and Xinjiang) aim to train and recruit 36,832 medical personnel for the region over the ten years following the war.[8] The group does not appear to have determined precisely how many medical personnel were in the region at the time of their survey, but they did look at medical education in the northwest and found that there were hospitals and schools training doctors, nurses, and midwives and that together, these institutions trained about 1,475 medical personnel per year.[9] Clearly, the supply was not adequate to meet the demand. Therefore, the group proposed strategies for training additional medical personnel during the postwar period. For water management, the report proposed a steadily escalating supply of technicians with varying levels of expertise. In the first year, they estimated a need for 30 high-level technicians who had graduated from universities or specialized training schools, 150 middle-level technicians, and 200 lower-level technicians who had at least a middle school education. They also projected that they might need ten additional people with various other kinds of expertise. The total water management personnel need that they estimated for the first year of the postwar plan, then, was 390. By the tenth and final year of the plan, however, the group estimated that the northwest would require a total of 6,970 water management technicians.[10] In the case of water management, as for medical professionals, estimated personnel needs were high, and existing educational systems were not capable of producing such high numbers of technicians.

Using the Education System to Develop Human Resources

These late-wartime plans for postwar development were not the first human resources development programs that the Nationalist government had devised. The Ministry of Education had already started work on a national plan for human resources development in the very early stages of the war and was at that time considering a reorientation of educational priorities to better align with national needs in a time of war. The Ministry of Education had greatest control over higher education and thus, starting at the Emergency National Congress in April 1938, focused its reform efforts in that quarter. The three main points of the ministry's wartime educational program were "(1) the formulation of an educational policy for the promotion of social welfare and interests; (2) the establishment of closer co-operation between educational and national reconstruction movements; (3) national defence and the training of technical experts."[11] To meet these goals, the Ministry of Education initiated a program to nationalize private universities and elevate technical schools to the level of universities so that by the end of the war, the number of state-controlled universities had expanded considerably. At the same time, they standardized curricula across all state-run institutions and expanded the number of fields of study available to students in science and technology, creating some programs that were completely new to Chinese higher education and expanding the availability of other programs that had previously only been offered by one or two institutions.[12] Table 2.1 shows that the new courses of study introduced in 1939 were all in applied science and technology, and in theory, at any rate, all of the graduates of these programs would be able to rapidly integrate themselves into the working world and

Table 2.1   New Fields of Study Introduced by the Ministry of Education in 1939

Institution	Southwest United University	Yunnan University	Wuhan University	Jinling University	National Central University	Northwest Agricultural College
New Fields	Telecommunications	Mining and Metallurgy	Applied Chemistry	Auto Engineering	Animal Husbandry and Veterinary Science	Agricultural Economics
Institution	Kweiyang Medical College	Chungking University	Fudan University	Central Political Institute	National Technical Institute	National Technical Institute
New Fields	Health Administration and Public Health	Statistics	Accounting	Accounting and Agricultural Economics	Agricultural Production, Paper Manufacturing, Tanning, Dyeing	Sericulture, Marine Product Processing

Source: Freyn, *Chinese Education in the War*, 95.

contribute in concrete ways to China's economic reconstruction. As Minister of Education Chen Lifu noted, "Special emphasis was laid upon courses with direct bearing upon Chinese problems, thus correcting the former tendency of neglecting them in the curriculum."[13] Before the war, the Ministry of Education had already attempted to correct this trend by requiring that all universities have a college devoted specifically to a science such as medicine, agriculture, or engineering, insisting that 70 percent of its subsidies to private institutions be dedicated to promoting science education, and mandating that more new students be admitted in the sciences than in the humanities and arts. The additional strategies that the ministry developed during the war accelerated a process that was already under way.[14]

In addition to encouraging the establishment of new departments that would provide the kinds of practical technical instruction that the Ministry of Education believed China needed, the ministry also sought to establish a set of educational standards for those and other fields. Beginning in the spring of 1938, the Ministry of Education started collecting both syllabi and suggestions from faculty at all institutions of higher education with the aim of standardizing curricula across all such institutions. The result was a set of curricula that focused on general education in the first year and funneled students into their major fields in the second year and more-specific subfields for the last two years.[15] It is difficult to perceive from the curricula themselves whether these revisions in and of themselves achieved the effect of making education better suited to meeting China's technical human resources needs, but what is clear is that the very purpose of curricular revisions was to bring about a centralization and standardization of university-level training across institutions governed by the Ministry of Education. All such institutions were to follow the standard curriculum. Therefore, in theory, at any rate, a physics student in one institution would take the same set of courses as a physics student at a different one and all physics students would emerge from their college programs with the same training. However, the plan did not take into account the possibility that instructional resources and faculty expertise might vary from institution to institution or that an expansion of the number of programs might strain existing instructional resources.

Motivated by concern over the Ministry of Education's approach, educators occasionally volunteered their own alternative plans for the improvement of education in their fields. "Realizing the importance of good quality of training for students of agriculture," a number of "prominent specialists in agriculture" met at the College of Agriculture and Forestry of Jinling University in Chengdu toward the end of 1940 and drafted "a request for the improvement of agricultural education" that was then signed by around forty leaders in the field and submitted to the Ministry of Education. The ministry, according to an article published in *Agriculture & Forestry Notes*, asked for details, in response to which Dean C. W. Chang of Jinling's College of Agriculture and Forestry submitted a detailed plan. The proposal divided China into eleven agricultural regions, each of which would be served by a state-run college of agriculture that would oversee lower-level agricultural schools (middle and vocational schools) in that region. Each college would have an agricultural education department to train teachers for secondary schools. Some regional colleges, if they met certain standards, would offer graduate instruction. All foreign experts advising the Nationalist government would be attached to a college of agriculture as an instructor. All extension work would be done by provincial bureaus of agricultural improvement, and not by the agriculture colleges, but research work would be divided between the two types of institutions, which would collaborate as closely as possible. The group proposing this structure was motivated in large part by what it saw as an excessive proliferation of colleges of agriculture that overtaxed the limited human resources in the field, leading to a dilution of any given institution's ability to perform to its potential. Their proposed plan aimed to create a rational system that would concentrate well-trained agricultural specialists in a smaller number of institutions, each of which would provide educational and research services to its region.[16]

Partly but not solely through the nationalization of universities, the number of institutions of higher education increased over the course of the war. Y. C. Yang 楊永清, president of Suchow University, noted in an address at the Southern University Conference in Atlanta in 1944, that the 108 institutions of higher education that had existed prior to the war had multiplied to 133 (this growth occurred in spite of the suspension of operations of seventeen institutions, meaning that forty-two new universities had come into being between 1938 and 1944) and "fields of education not covered before are now taken care of and regions not yet served by colleges and universities before now have such institutions established. Not only in number of institutions, but also in the number of students there has been a steady and encouraging increase. Against 42,000 college students in 1937, there were in 1943 a total of 57,000 college students, an increase of over 35%."[17] From Yang's perspective, these changes, including the opening of universities in China's western provinces, were all a positive sign that access to education was increasing.

That said, such rapid expansion was not without problems. As C. W. Chang's proposal for reform of agricultural education highlighted, there simply were not enough faculty to staff all of the new colleges of agriculture. Wilma Fairbank, one of the more vocal critics of the Kuomintang's (KMT's) wartime efforts at educational reform, concurred, later observing that under Chen Lifu, "the number of schools and universities was expanded during World War II far beyond the availability of qualified teaching staff or equipment. Consequently, standards were drastically reduced."[18] In addition, under wartime conditions, funding was insufficient, and the Ministry of Education, according to Fairbank's 1948 report on Chinese education for the United Nations Educational, Scientific and

Cultural Organization, received no more than 4 percent of the national budget, not because education was not prioritized, but because the bulk of China's budget was going first to war against Japan and then to civil war.[19] Although the Ministry of Education succeeded in expanding the number of scientific and technical programs available to students, observers, including instructors in those programs, questioned the extent to which they were able to deliver a high, or even good, quality of education.

This was not the first strategy the Ministry of Education had employed to grow the number of students in middle schools and universities. In the early 1930s, to increase the number of middle school and university students, the KMT introduced legislation that shielded both students and graduates from the draft. The KMT's first conscription law, promulgated in 1933, exempted "'all middle school graduates or those possessing equivalent qualifications' from enlistment in the army."[20] As Hubert Freyn noted, "The purpose of exempting the students from direct army service is to preserve them for the country's future. In proportion to the entire population their number is so small that war casualties of any magnitude would seriously threaten that so vital process of reconstruction the accomplishment of which is to China's leaders as important as the winning of the war."[21] By the early 1940s, however, the immediate military need for certain skills that only such students would have was beginning to outweigh the need to incentivize students to enroll in school. As a result, some exceptions were being made, as "educated youth were badly needed by the military."[22] By 1941, students in nursing, education, and engineering were frequently required to commit to a period of government service as a condition of matriculating in a course of study. According to Chen Lifu, medical and engineering students were put to work in military hospitals and at construction sites for new military airports and foreign language students were called upon to serve as interpreters for American pilots and other military personnel. In 1944, for example, 3,267 students were serving as interpreters and 3,104 students were serving in medical fields and engineering.[23]

The Ministry of Education was not simply interested in expanding the number of students it reached. It also sought to direct those students to pursue training in areas of need. To that end, it introduced entrance examinations in 1938 with an eye to placing more students into scientific and technical fields at the university level, although it is unclear whether this strategy made any difference in the early years. According to the China Information Committee (a government publishing organ), "In 1940, 3,771 out of 7,024 students who passed the national matriculation tests took engineering, science, agriculture or medicine as major courses, while 2,443 specialized in arts, law and commerce and 810 others were to be trained as teachers. In 1938, when the matriculation tests were introduced, 2,942 out of 5,460 qualified applicants majored in sciences, while 1,427 chose arts and 1,091 selected teachers' courses as their chief studies."[24] In other words, although the absolute number of students matriculating in the sciences in China's institutions of higher education increased by 829 between 1938 and 1940, the percentage of students matriculating in the sciences did not—in 1938 it was 53.8 percent, and in 1940 it was 53.6 percent. The efficacy of the entrance examinations in achieving their intended goals, therefore, is unclear, and it appears that at first, they did little to steer students toward the sciences. If one factors in vocational and technical schools, however, the numbers look quite different, and there were large jumps from 1935 to 1945 in engineering, medical fields, and agriculture (see table 2.2).

In 1940, the Ministry of Education and the NRC devised a new and specific plan to encourage students to matriculate in scientific and technical fields and to encourage professionals, especially academic professionals, to focus more of their research energy on producing the specific kinds of scientific and technical knowledge that China needed. This strategy, the National Defense Science Movement (NDSM, 國防科學運動), was overseen by the new Science and Technology Planning Committee, which was placed under the authority of the Executive Yuan. The movement had three main goals. First, it aimed to educate the people in the importance of building national defense industries. Second, to expand the capacity of such industries, it aimed to promote research with industrial applications. Third, it aimed to improve and expand science education at all levels so as to increase the quantity of scientific personnel that China would have available for postwar reconstruction. To achieve these ends, the NDSM initiated a mass education campaign, laid out a plan to push more students into scientific and technical fields of study, increased funding to so-called national defense institutions, and sent technicians abroad to familiarize themselves with industrial practices in the United States (a topic that will be dealt with in the next chapter).[25]

Mass Education

The idea of popularizing scientific and industrial knowledge was by no means new during the wartime. Already in 1932, Chen Lifu was trying to use the resources of the state to popularize science. The methods he employed—creation of a Chinese Association for the Scientization Movement (中國科學化運動協會), organizing exhibits and radio lectures, and publishing magazines—were quite similar to the strategies implemented by the NDSM. But popularization of science was not solely the realm of the state, nor did it start after the KMT came to power. Scientific and technical subjects were discussed as early as the late nineteenth century in pictorial publications and in most of the more popular new journals of the new culture era. Other popular journals such as *Women's World* and *Ladies' Journal* included information on the industrial arts. Journals that focused specifically on scientific topics did not come into being until the early 1930s. In the same period, modern museums and exhibition halls also started hosting exhibits with scientific and technical content.[26]

What was new about the NDSM was that it explicitly linked science to national defense and that it expanded the role that government should play in encouraging popular interest in science. To popularize the idea of national defense science, the Ministry of Education asked county- and city-level governments to set up local science organs to promote science in elementary and middle schools, establishing rules for their formation in 1941.[27] By 1943, many such organs had indeed been established and they were reporting regularly on their activities to the Ministry of Education. For example, one county-level propaganda committee for the Movement to Promote Science (科學化運動), a part of the NDSM, reported that they had decided to do a weeklong set of science activities such as performances and expositions.[28]

These provincial science institutes and local science halls served, for the most part, as crosses between museums, libraries, and small-scale factories where scientific apparatuses, models, and other instructional materials for use in schools were produced, people could come to reading rooms or simply stop by and look at the posters set up outside to learn scientific information, and exhibits about the scientific features of the region were displayed. They also did science broadcasts over the radio (though it is unclear how many people would have heard them). As British scientist Joseph Needham wrote in his notes on the Gansu Science Education Institute (甘肅科學教育館), they were making tuning forks and pulleys out of aluminum salvaged from Japanese planes and brass weights made from old bronze coins. Both Dorothy and Joseph Needham were particularly impressed with an exhibit on insect wax that they saw at the Guizhou Provincial Industrial Products Museum (貴州物產陳列館), located in Guiyang. The Popular Science Hall (四川省立科學館) in Chengdu offered a meeting place for science teachers and had a taxidermy exhibition and a garden with geometrical shapes.[29]

Organizations such as the Gansu Science Education Institute, which had been created with Boxer Indemnity funds in 1936 by J. B. Taylor and Y. P. Mei 梅貽寶 of Yenching University, a private university not under KMT control, engaged in a variety of activities aimed at improving the scientific environment in Lanzhou. Needham, visiting the institute in 1944, found that, like other institutes he had earlier come across in Yunnan and Sichuan, it was active in making scientific apparatus to be used by regional schools as well as boxed sets of specimens of things like ore. Perhaps even more important from a mass education perspective was that the popular science division of the institute published a wall newspaper every ten days that was posted at the Lanzhou civic center for the public to view (see fig. 2.1). "During my ten days at Lanchow, I noted interesting articles on Archimedes and the history of geometry, on twinning and other questions in experimental morphology, and on parasitic insects."[30] Although fascinating to Needham, these topics might not have been particularly accessible to a nonscientist, and consequently, the Gansu Science Education Institute may not have been very effective in popularizing science. Needham offers no indication of whether there were throngs of people reading these newspapers, and a photograph he took shows the newspaper board alone with no potential readers in sight.[31] But even if people did not, or could not, avail themselves of all of the educational opportunities afforded by local science organs, those organs did at least attempt to provide such opportunities.

The Ministry of Education may have sensed this distance between government science organs and the masses because by 1945, it was requiring all educational institutions at all levels to hold science activities on national holidays such as National Day, Children's Day, and Youth Day.

FIG. 2.1. A publication on popular science displayed outside the Gansu Science Education Institute. Courtesy of Needham Research Institute.

Schools were asked to show films about science, host educational lectures, hold speech contests with science themes, and hold open houses in laboratories and libraries. The idea was both to energize young students and draw nonstudents into schools so that they might learn more about scientific and technical activities going on in China and elsewhere and, perhaps, find inspiration to develop their own technical expertise.[32]

Chances are high that many of these activities took the form of educational films, demonstrations, and presentations rather than the sorts of interactive, hands-on activities that might truly have spoken to people who were not already engaged in academic study. Nonetheless, some displays did succeed in drawing large audiences. One of them was the March 1944 Chongqing Industrial and Mining Exhibition (see fig. 2.2), which "occupied large and spacious halls, specially constructed for the occasion entirely of bamboo poles and matting and set up on the campus of the Chiuching Middle School in pleasant surroundings."[33] According to Needham, who wrote an article on the exhibition for the journal *Nature*, it attracted many thousands of visitors each day. A different article claimed that it "was the biggest and the most successful exhibition

FIG. 2.2. Chiang Kaishek, Weng Wenhao, and others inspecting the oilfield exhibit
of the Chongqing Industrial and Mining Exhibition. Courtesy of Needham
Research Institute.

ever organized in China" and had been attended by two hundred thousand people.[34] The exhibits were on mining and metallurgy, the chemical industry, and engineering and included displays of Chinese-made
chemical products made of vegetable oils and oil from the newly developed Yumen oil field, such as candles, vanishing cream, paint, and even
petrol. Music for the exhibition was played by Chinese-made record players, topographic models with running water demonstrated how rivers
could be dammed and harnessed for hydroelectric power; other displays
contained miniature working models of new kinds of machinery, but large
diesel engines manufactured in China were also on display.

Needham walked away from the exhibit convinced that it demonstrated "that given the tools of the trade, Chinese technologists, engineers
and scientific men are the equal of any in the world," which surely was
one of the purposes of the exhibit.[35] In addition to building precisely the

kind of enthusiasm for Chinese scientific development that it aroused in Needham, the intent of this exhibit was to demonstrate to students and the public the various ways in which science and technology could be used to defend, save, and reconstruct the nation.[36] A third intent was to encourage students to embark on careers in science or to get technical training so that they would be able to participate in postwar reconstruction efforts. One nine-year-old attendee, who went to the exhibition with his family and went on to study physics and meteorology, later remembered being especially impressed by the electric mining model and other toy-like models and suggested in his memoir that the exhibit actually did play a role in his later decision to go into the sciences and become a meteorologist.[37] Needham felt that the exhibits were of such a high quality that they should form the basis of a national museum of science and industry in the postwar period, and in fact, part of the exhibit on standardization did end up in the West China Science Museum in Beibei.[38]

Large-Scale Training Plans

One objective of the NDSM and the Science and Technology Planning Committee was to dramatically increase the number of skilled technicians who would be able to contribute to postwar reconstruction. To this end, human resources surveys were undertaken and training plans were developed. The "Resolution to Promote the National Defense Science Movement" laid out a plan to train thirty-nine thousand engineers and eleven thousand natural scientists in the decade between 1941 and 1951. With these numbers, state planners believed that China's industrial development prospects would greatly improve. In addition to higher education, the plan targeted vocational education and short-term training, through which it estimated that it could provide scientific and technical training to a total of 89,662 people over a ten-year period, most of them students in vocational schools.[39] By 1942, the Science and Technology Planning Committee was conducting surveys of China's existing pool of scientific and technical talent, investigating how scientists in other countries were mobilized and distributed, and coming up with strategies for training and utilizing China's scientific and technical talent.[40] Not long

Table 2.2   Number of Students in the Sciences at the Vocational Level and
Above

Year	Total Number of Students	Physical	Engineering Sciences	Medicine	Agriculture
1932	42,710	4,159	4,439	1,852	1,557
1935	41,128	6,336	5,514	2,977	2,163
1940	52,376	6,090	11,227	4,271	3,675
1945	83,498	6,480	15,200	6,291	6,380
1947	155,036	10,060	27,579	11,855	10,179

The huge jump in numbers between 1945 and 1947 is due to the inclusion of students in previously occupied areas. Worth noting is that the number of science students in college did not increase quite as dramatically as the numbers in the chart above. Vocational education in the sciences was rapidly expanded during the 1940s and some of the growth in science education was taking place at that level. This table was originally published in Greene, "Looking Toward the Future," 290. Reproduced by permission of the Institute of East Asian Studies, University of California, Berkeley.

Source: *The Second China Education Yearbook*, 1403, 1412.

after that, the ten-year human resources projection figures were substan-
tially refined so as to include projections for virtually every technical field
that might be of interest to the Ministries of Defense, Finance, Communications, Economy, Agriculture and Forestry, and the NRC. Although
it is difficult to assess the impact or success of a movement like the NDSM,
and many other factors may have played a role, as table 2.2 shows, numbers of students in the sciences, especially in vocational education, did
increase during the wartime period.

Vocational Education

Vocational education had found support in China among Chinese businessmen as well as from some of China's leading educators in the late
1910s, who advocated for its development through the Chinese Society
for Vocational Education (中華職業教育社) and oversaw the establishment
of Shanghai's first vocational education school. That school offered courses
in basic science, Chinese and foreign languages, business, and manufacturing of specific products and provided internships at various Shanghai
banks and enterprises.[41] In the 1930s, the Ministry of Education had taken
up the idea of vocational education and also started to establish vocational

schools. During the war, a number of different types of vocational education programs were developed in China. In addition to conventional vocational schools, other government and nongovernment organs established vocational education programs with the idea of creating pathways for students to develop the skills and knowledge required to implement the economic development goals of those organs.

One ministry that was particularly active with respect to development of vocational training programs was the MOAF, which was a central government organ but worked closely with provincial and sometimes even county-level partner organizations.[42] Under the auspices of the MOAF, but not in a fully planned and centrally organized way, a variety of training programs were developed during the war that aimed to produce the kinds of skilled personnel that agricultural leaders, at both the central and local levels, believed were needed to further new agricultural development strategies.

One example of such a strategy is the Liuchow Agricultural Vocational Middle School (廣西高級農業職業學校/柳州高级农业职业学校). The school was attached to the National Agricultural Research Bureau (NARB) Experimental Farm that was maintained by a group of national and provincial entities. The school was started in 1940 by Guangxi native and Cornell PhD Ma Baozhi 馬保之. During the war, Ma was sent by the NARB to Liuzhou to direct its experimental farm in Guangxi. While there, he also served as the head of the Guangxi Provincial Experimental Station and head of the MOAF's Southwest Unit for the Promotion of Animal Breeding and Training (西南各省推廣養殖站及推廣訓練班). Training, therefore, was understood by the MOAF to be one of Ma's responsibilities. He established the school to give students an opportunity to undertake practical training in a real-life setting.[43]

As Joseph Needham learned during a 1944 tour of the facility, the school was intended "to train extension workers for remote country hsien, so as to bring to them a knowledge of modern scientific agriculture and rural technology." The students lived in villages that consisted of twenty-five students plus a teacher in which they were entirely self-sufficient. They performed two hours a day of agricultural labor and took classes in agriculture, forestry, animal husbandry, and veterinary science, using their villages as experimental laboratories where they raised pigs and cattle and planted crops. After three years of study, most of the students appear to have gone on to do actual agriculture extension work. Very few of them went on to college, which was precisely the intended result. The school's first class had seventy students, both male and female, all middle school graduates. By 1942, the student body had increased to 266 students. As Needham wrote in his report on his southwest trip, "This school is one of the most hopeful things I have seen in China."[44] In fact, the school was not just a temporary institution. It has continued to exist with fundamentally the same mission and is now known as the Guangxi Eco-engineering Vocational and Technical College.

Like the NARB's Guangxi branch, Jinling University's College of Agriculture and Forestry also established a pair of schools to educate farmers in 1938. The two schools focused on farming practice with courses that were taught entirely by extension workers. The college's Farmers' Foundation School was located in a village about twelve miles from the Jinling University campus in Chengdu, and by 1940, it had about seventy students who were divided into two classes by age. The older students attended lectures and demonstrations that were "based on the needs of the community" for a year and worked to translate what they were learning into daily practice, whereas younger adults whose labor was more heavily depended on would limit their studies to a six-month-long low-season course during which they would attend class for two hours per night. The Farmers' Elementary School was located in a different county and ran classes from November to March, expecting students to attend for two years. The students, who in the winter of 1940 numbered about forty, did classwork in the morning and fieldwork in the afternoon. Neither school went so far as to attempt to simulate village conditions by housing students in their own experimental villages, but the aims of the Jinling University Farmers' Schools were nonetheless similar to those of the Liuzhou Agricultural Vocational Middle School. The Liuzhou school aimed to prepare semiprofessional extension workers who would live among farmers and offer advice on a wide range of farming strategies. The students in the Farmers' Schools, themselves active farmers, worked the land by day and studied by night, or were the children of farmers. Nonetheless, they were expected to do some extension work and were required to "grow improved seeds and use improved methods in their farm work" with the expectation that they would gradually "become key farmers in the community for the improvement of agriculture."[45]

Not all vocational training was done in the context of schools, however. In some fields, training was overseen by the government's Bureau for the Training of Skilled Workers, which was specifically focused on training workers for arsenals as well as both state-owned and private factories. To that end, the Ministry of Economic Affairs (MEA) reported in 1944 that it had apprenticed five hundred workers in arsenals for three-year training periods, one thousand ordinary skilled factory workers in state-owned factories for two-year training periods, and three hundred apprentices per year for one-year training periods in private factories. Over the course of 1944, according to the MEA, 3,600 workers graduated from these training programs.[46]

Training College Graduates for Practical Work

In many fields, however, the more education people received, the less likely they became to actually labor with their hands and the more removed they became from real conditions in rural and industrial China. How to bridge this gap between the educational system and China's need for technicians who would engage in hands-on work was a problem that occupied the minds of virtually everyone who tried to develop and manage training and modernization projects in China during the war. Many graduates needed to be taught to work with machines or the land, and they needed experience working in real-life situations.

The story of the Agricultural Bureau (農本局) well illustrates this problem and one training approach that was developed to address it. The Agricultural Bureau was set up to facilitate agricultural credit and promote marketing of agricultural product. Soon after being appointed vice-minister of economic affairs in February 1938, Franklin Ho began concentrating his efforts on developing this institution so that it could effectively accomplish its goals. Ho set up local banks, irrigation projects, and seed improvement projects and hired a host of recent college graduates, each of whom would work simultaneously as banker, extension agent, and warehouse manager in the county-level Agricultural Bureau office to which he was assigned. Recognizing that there were few people in China who were actually trained to do this kind of work, within his first month on the job, Ho set up a training institute in Chongqing for the college graduates he recruited, which he ran along with the economist H. D. Fong 方顯定. Attendees typically underwent three months of training. In its first two years, Ho's institute taught six hundred to seven hundred trainees, providing them with "lectures on the practical aspects of accounting, business, cooperatives, rural credit, etc., delivered by the heads of the various departments" of the Agricultural Bureau. Ho's training program was thus initially focused entirely on introducing trainees to the technical knowledge that pertained directly to the job. With time, however, Ho discovered that technical knowledge alone was not sufficient to prepare the trainees for the challenges of their positions. To address this problem, "we added lectures designed to acclimate the trainees to living and working in rural areas. Conditions out in the field were primitive, and trainees had to be introduced or reintroduced, as the case might be, to elementary first aid, sanitary, and health measures." Ho and Fong also began giving "pep talk"–type lectures that "encouraged the trainees [*sic*] and informed him of how important his work was," and in so doing helped to develop a strong sense of mission and commitment to the institution among the trainees.[47]

Ho's trainees were nearly all recent college graduates who had migrated west in the first months of the war and were eager to find work. Although they were well educated, they had neither adequate technical knowledge nor life skills to undertake the challenging work of the Agricultural Bureau. In particular, as Ho learned by sending these young workers out into the field, they were generally ill-prepared to work in small rural villages with uneducated farmers.[48] To a certain extent, therefore, the training that Ho developed for them must have been something akin to the kind of administrative and technical training that a low-level European colonial administrator might have undergone before heading off to the colonies but with one major exception: Ho expected his trainees to engage directly with the locals, rather than remain aloof, and he therefore had to adapt his training program to meet those needs. Although Westerners and Chinese alike complained that many educated Chinese did not like to "get their hands dirty," Ho clearly expected that his employees would do just that, and the fact that they did it suggests that this trope about effete Chinese intellectuals had little basis in reality. However, he also concluded after observing his workers in the field that many of the people he hired needed to be taught how to do it and how to safely operate in an environment that was distinctly less modern than the environments from which they were coming.

Because the workers of the Agricultural Bureau were dispersed across western China and worked in villages on their own or with one other colleague, "the success of the work [of the entire institution] depended heavily on the loyalty, devotion, dedication, and training of the personnel."[49] From Ho's perspective, the program was very successful. It organized nearly three hundred cooperative banks and/or warehouses between 1938 and 1940, developed targeted economic programs that matched the agricultural particularities of the counties they were created in, and also had considerable success multiplying new seed that promised to thrive in the region.[50] In addition, Ho felt the program was successful in training good people who understood the role they needed to play. One pair of Agricultural Bureau workers he met on a site visit were graduates of Jinling and Tsinghua Universities. "Both of them had gone through a short period of training in the institute . . . in Chungking before they had been sent out into the field. That morning they had themselves gone to the countryside, into a village, to carry out what might be called public relations and education work. They explained that although they had several subordinates (it was the practice to recruit on the spot several people from the hsien [county] who were familiar with the countryside, but normally these were people with middle school training at most), they felt the subordinates could not successfully carry on the actual business in the village. Therefore, every day they went out themselves. They said that there was not much business in the office, that one had to go out to the fields to accomplish anything and to find out just what the needs and problems of the peasants were."[51] These men were doing precisely what Ho hoped they would do—applying their advanced knowledge and skills to the concrete problems of the farmers in their area and engaging in practical hands-on work when they saw that it was needed. Well before the war had ended, the Agricultural Bureau had been asked to take on more work than it could handle and was then reorganized, and by 1941, Ho had left it. Whether this training program had a lasting impact in the countryside is unclear, but it seems likely that it at least had a lasting impact on the hundreds of Agricultural Bureau employees who were trained in its program.

Nongovernment Training— *China Industrial Cooperatives*

The CIC (工業合作社) was another entity engaged in vocational training, but its approach was dramatically different from that of the Agriculture Bureau. Whereas Franklin Ho took educated young men and trained them for life in the countryside, the CIC took uneducated youth and peasants who showed promise and trained them to do technical jobs. On the one hand, government programs tended, by and large, to focus on production of skilled technicians and administrators, and even government vocational programs provided students with more education than they probably really needed to function as skilled laborers. On the other hand, the CIC was focused on building a workforce from the ground up and built programs that aimed specifically at turning unskilled labor into skilled labor.

The CIC was founded in Shanghai in 1937 by a group consisting largely of left-leaning Western expatriates, including Rewi Alley, Edgar Snow, Nym Wales, and Ida Pruitt, and some Chinese, including communists Hu Yuzhi 胡愈之 and Sha Qianli 沙千里, all of whom enthusiastically endorsed a bottom-up approach to social and economic reconstruction based on the empowerment of impoverished, rural Chinese. Following the Nationalist retreat to Hankow, the CIC started to receive some support from the Nationalist government, but it continued to receive support from and operate under the auspices of INDUSCO (International Committee for the Promotion of Chinese Industrial Cooperatives) and thus, unlike the other institutions mentioned above, it was not a government program. Some of its vocational training took place in factories and some took place in so-called Bailie Schools in Shuangshipu and Lanzhou, both in Gansu Province, and in Chengdu in Sichuan Province. Joseph Bailie, a missionary and teacher after whom the schools were named, had emphasized the importance of practical training for both unskilled Chinese laborers and Chinese university students in fields such as engineering.[52] The first two Bailie Schools were founded in Ganxian (Jiangxi) and Baoqi (Shaanxi) in 1940, not so much as schools but as, in the case of the Ganxian school, a training course to prepare workers for the CIC's machine shop in the same location, and in the case of Baoqi, a sort of experimental factory. In April 1941, the CIC opened its third and eventually most successful Bailie School in Shuangshipu at which a generalizable half-class, half-work curriculum founded on the principles of work, technical skill, and cooperation was gradually developed, though the school was hampered by various challenges, including staff problems. The Lanzhou Bailie School, which opened in 1943, was the first school to be properly planned by trained educators and was in many ways the most successful of the schools, though it too suffered from staffing and equipment challenges. Bailie Schools were also opened in Chengdu and Chongqing, both focusing educational efforts on specific skills that could not be developed in the other schools owing to lack of access to equipment.[53] In December 1944, concerned about possible Japanese offensives as well as increasing threats that the students would be conscripted, George Hogg, head of the Shuangshipu school, and Rewi Alley, who oversaw the Bailie School project, removed that school to Shandan, in a considerably more remote part of Gansu Province, where they reestablished it in an old temple. By the time the CIC set up the Shandan Bailie School in the spring of 1945, Hogg and Alley had ample experience of what worked and what did not and were able to create a much more successful venture.

The Shuangshipu Bailie School was established in 1941 with a student population made up of war orphans and local peasant children and sought to train them in skills that were both marketable and useful with an eye to contributing to the development of the local economy. The idea behind the school was that "ordinary schools would not produce the necessary people." In other words, the skills and knowledge that were taught in regular schools did not match up to the skills and knowledge that farmers and workers needed to be able to get ahead and develop their own local economies.[54] In the view of the CIC leadership, regular students simply were not equipped to meet the challenges of the moment. As Hogg noted in a June 1942 letter to Ida Pruitt, "Everybody I meet falls for the idea that we must get hold of the young worker and technician type, who are China's real hope. The student type is getting more and more useless as conditions get tougher. They are the first to fall into the 'every man for himself, it's no use trying to do anything' attitude."[55] CIC leaders, thus, shared some of the frustrations of government education reformers, but they took a very different approach from that of either the Ministry of Education or Franklin Ho to resolving the problems they saw.

In particular, the Bailie Schools aimed to produce junior technicians who could bridge the gap between the CIC's trained engineers and the semiskilled workers in the CIC factories. These junior technicians could "on the one hand, understand the mental attitudes and native dialects of the coop workers, and on the other hand follow the ideas (spoken, written, or drawn in blueprint form) of senior technicians and engineers."[56] They would be able to understand the vocabularies and cultures of both groups and would serve as lynchpins in the CIC's factories.

Not only did CIC leaders have very particular ideas about whom they should train and for what purpose, but they were also very concerned about the type of education they would offer and who would be most appropriate to deliver the instruction. They were particularly careful to avoid hiring highly educated Chinese men of letters, with whom they had had bad luck because the expectations of the ones they had previously hired did not correspond well to the realities of the schools. A 1942 proposal for the new Bailie School in Lanzhou argued that "care must be taken with regard to the selection of staff, and teachers. These had best not be educational specialists, unless those of an exceptional kind are obtainable." From the perspective of the Western leaders of the Bailie Schools, who had had some experience in trying to hire suitable Chinese teachers, "The average educationalist (in China rather apt to be a textbook purist) will throw up his hands in horror at the Bailie School, which falls outside all previous categories—primary, middle, or technical. After struggling for a week or two to fit the students of mixed ages, backgrounds and academic standards into his preconceived pidgeon-holes, he will spend the last of the school's administrative fees in sending three polite telegrams of resignation, and pack up. Thus the Shuangshihpu Bailie School changed headmasters eight times in 18 months, with the boys getting more exasperated and less easy to educate each time."[57] The Shuangshipu school, according to a 1942 report, was especially eager to find students "who will bring an understanding of local conditions and problems as well as dialects" so that they might serve to "bridge the gap between the highly trained city-educated engineers and the cooperative member workers." Students were grouped by occupation, and the 1942 Shuangshipu students were schooled in the fields of textiles, mechanics, truck motor mainte-

FIG. 2.3. Two boys from the Tibetan borderlands learning about screw threads at the Bailie School in Lanzhou. Courtesy of Needham Research Institute.

nance, and accounting, but with the hope that types of instruction would improve and expand with time.[58] Some of the instruction at the Bailie Schools involved development of traditional handicraft skills, such as weaving, brick making, and leather tanning, but students received instruction in more modern technologies as well (see fig. 2.3). Initially set up in old temples contributed by the local government, the schools gradually constructed new buildings.[59]

Students were drawn primarily from refugee populations so that the student body at Shuangshipu in 1942, for example, drew its fifty-nine students from nine different provinces. Of those students, forty were from seven Japanese-occupied provinces, and the remaining nineteen were from especially poor families in unoccupied Gansu and Shaanxi. Hogg described a number of the refugee boys from Manchuria, Anhui, and Jiangxi as being from middle-class families with some educational background.[60]

Among the staff at the Shuangshipu school, in addition to Hogg, were Ting Ch'i-shen, a twenty-year-old graduate from the local higher technical school who taught mechanics and supervised the machine shop, a couple of volunteers from the Friends' Ambulance Unit, one Chinese and the other Scottish, and K'ang Cheng-shung, a forty-one-year-old graduate of the Hebei Provincial Industrial School, who although able to teach courses on textile manufacture, was, according to Hogg, so bothered by the students' lack of understanding of "the customs of education" that he left the school. Of these men, only K'ang and Hogg had any educational background, and from Hogg's perspective, K'ang was too wedded to educational traditions to be comfortable in an alternative school such as the Shuangshipu Bailie School.[61] By 1944, things were looking up with regard to the staff at Shuangshipu and other Bailie Schools, as various Bailie School graduates were taking on instructional roles. Hogg mentions a young man by the name of Ting who had become "a first rate Bailie teacher, with a heart for the job, and incredible success in teaching machine drawing to so-called 'half-witted' peasant boys."[62] Another three graduates of the Shuangshipu school were teaching evening courses on leather work at the Lanzhou school while working by day in the Lanzhou CIC tanning cooperative.[63]

Most notable in Hogg's 1942 report on the Shuangshipu Bailie School is a three-page report on a trip to the Shuangshipu CIC coal mine by two Bailie boys aged thirteen and fifteen that details access paths to the coal mine, production and capacity of the mine, the types of industries they saw on the way, and the potential for other industrial development in the region. In addition, the boys remarked on numerous conversations they had had with local peasants about topics ranging from industry to military conscription.[64] The boys had taken what they had learned in the school and applied it, and Hogg was clearly proud of the report they had produced.

By 1947, the Shandan school (the successor to the Shuangshipu school) had 170 students, most of them from Gansu and Henan, and was expecting another group of sixty students from the Lanzhou Bailie School, which was about to close. At the end of the war, the CIC orphanage at Baoqi sent its children to Shandan, among whom were the school's first eight female students. The school was staffed by both Chinese and Western instructors who taught classes on subjects ranging from economic geography and geology to blacksmithing, weaving, and pottery.[65] Students were themselves responsible for food preparation, cleanliness, and maintenance of dormitories, as well as producing news articles for themselves.[66] By 1947, students were learning how to work with cotton, textiles, and leather; they were blowing glass, making pottery, paper, and water wheels; and others learned how to maintain and operate heavy machinery, procure materials, manufacture things with odds and ends, and farm.

The Bailie Schools in Shandan, Lanzhou, and Chengdu were but one part of the larger Gong He (工合 or 中國工業合作協會) movement, which also established cooperative industries in villages and sought to train adults for labor and leadership through those cooperative factories. The schools, since they involved training young people, were the piece about which there was greatest optimism. However, by 1947, Rewi Alley, head of the CIC in China, looked back at the Bailie Schools with a critical eye, concerned that their training efforts had not helped to expand and solidify the larger Gong He movement. In particular, he worried that trainees had been drawn from the wrong populations—displaced youth, the excessively young, and others who were there simply because it was a job—and that they had lacked sufficient depth "to turn out a man who would be properly equipped to undertake the very difficult work of village reconstruction."[67] Alley's greatest concern seems to have been the CIC's inability to foster in trainees the kind of commitment to both group and village that expansion of industrial cooperatives required. In other words, the bottom-up approach espoused by the CIC was not entirely successful in producing the kind of personnel that the CIC believed was needed to facilitate modernization at the village level.

Seen from a different perspective, however, Alley's complaints reveal that the program did have a certain amount of success. His trainees were abandoning the villages and making their way to the big cities on the coasts where "a university training is hoped for by all who can manage it, to be followed by trips abroad." Whereas trainable personnel was relatively abundant during the war, Alley's postwar challenge was "how to get people who will spend their lives working" in the "poverty-stricken villages which make up the bulk of China."[68] Alley and the CIC board were hoping to train lifelong commitment to both a movement and a locality into their students, but that did not always happen.

Other observers, particularly those with a leftist bent who viewed bottom-up cooperative movements with a favorable eye, saw great promise in the training program offered by the Bailie Schools during the war. Joseph Needham visited the Shuangshipu school on his northwest tour in the summer and fall of 1943 and came away with a very positive impression of the training efforts of the school. "The Bailie Schools," he wrote, "have impressed me as among the best examples of technical education which I have seen in China, ranking equally with the Agricultural Vocational (Extension Teachers') Training School at Shatang near Liuchow."[69] Needham was particularly impressed by the school's machine and textile shops, which provided on-the-job vocational training for the students, noting in his diary that the school would make a good model for other vocational schools in China.[70] The Bailie Schools, Needham wrote to an official in the British Foreign Office in 1947, "combined technological, general, and physical education, in just the right proportions to turn out the kind of men China needs almost more than any others, namely men of good technical training but also capable of democratic leadership in villages and small towns for the advancement of small cooperative industry."[71] Needham visited hundreds of schools, research institutes, and factories, writing up in his notes a catalog of scientific and technical activities in the unoccupied areas and was better positioned than anyone else to provide a comparative analysis of the qualities, positive or negative, of any of those enterprises. Even so, his enthusiasm for both the Bailie project and the Liuzhou school was undoubtedly also partly a consequence of his Marxist political leanings and his sympathy for cooperative movements in general, and it is quite possible that these sympathies led him to overlook the schools' flaws.

Bringing in Foreign Experts

The CIC and the Bailie Schools represent one way in which foreign expertise was utilized in China during the war. The CIC was managed by an international board led by the old China hand Ida Pruitt that operated out of the United States. Its founders were mostly Westerners, all left-leaning China hands with considerable in-country experience and connections.[72] During the war, the organization's main leader in China was Rewi Alley, a New Zealander. Throughout the war, the CIC lobbied for and received financial assistance from the U.S., British, and Chinese governments. In spite of these connections to government, however, CIC activities were really guided by the organization's own leadership and aimed at developing human and natural resources at a very local level. It may have been a foreign-led organization, but CIC's aims, as we have already seen, were very much to produce a bottom-up type of development, albeit one that conformed to ideas the organization's foreign leaders felt would be best suited to meet China's needs.[73]

This type of externally managed philanthropic structure was one model of wartime technical training involving foreigners, but it was not the only one. The Nationalist government also worked directly with foreign governments to organize other kinds of training opportunities that more clearly matched its own perceived needs, although it was unquestionably the case that those governments sought to make sure that the training served their own purposes as well. Both Britain and the United States sent foreign experts to China during the war to aid with various development projects, and many of the projects the U.S. experts worked on had a clear technical training dimension.

About eight months before the United States entered the war, the U.S. State Department's Division of Cultural Relations began working on a cultural relations program to provide assistance to China. The U.S. State Department established its Division of Cultural Relations in 1938 "for the purpose of encouraging and strengthening cultural relations and intellectual cooperation between the United States and other countries."[74] Its activities were confined to Latin America until it established the cultural relations program in 1941. Limited in scope and ambition by its budget, the program initially aimed to amplify existing initiatives that had been undertaken by private entities and thus engaged primarily in cultural exchange directed at elites. In the fall of 1941, however, Vice-President Henry Wallace directed it to start doing work that would improve the economic well-being of nonelites in partner countries.

The decision to extend the program to China came in the spring of 1941, at about the same time as the U.S. government extended lend-lease assistance to China. Over the summer and fall of that year, the advisory committee of the Division of Cultural Relations debated how best to enact the program in China. According to Fairbank, the committee leaned toward pursuing the same sorts of high culture programming that had been extended to Latin America, but Laughlin Currie, administrative assistant to President Roosevelt, was insistent that assistance should be given in civilian technical fields, including management, finance, and public administration. Following the attack on Pearl Harbor, Currie's approach won out. When the program was finally proposed to the Chinese government in early 1942, the emphasis was primarily on providing assistance that could "be of use to China in raising its standard of living, improving the condition of its rural population, assisting in the development of educational, social and administrative programs and thus contributing to China's war effort."[75]

By March 1942, the two parties agreed to a program through which the United States would send to China technical experts in a variety of fields requested by the Chinese government, and the Chinese government had provided U.S. Ambassador Gauss in Chongqing with an initial list of fields in which it sought assistance.[76] Between April and July of that year, the two governments hammered out the details of the program. The Chinese government was particularly eager to bring to China scientists and engineers specializing in potato farming, soil erosion, animal husbandry, chemical, mechanical, aeronautical, and electrical engineering, petroleum refining, shipbuilding, physics, and journalism.[77] At first, requests were funneled through the embassy, but with time, individual government organs started to make personnel requests directly to the Division of Cultural Relations. Academia Sinica's Institute of Medicine, for example, made a June 1945 request for a blood plasma expert to help them improve plasma production and advise on blood bank operation, and laboratory diagnostics.[78] Similarly, the NBIR also wrote directly to the Division of Cultural Relations in 1944 to request technical literature on the ceramic, plastic, leather, cellulose, oil and fat, pure chemical, and electrical metering and measurement industries.[79] By mid-July 1942, 175 American candidates had been asked to submit records, and 100 had been interviewed to determine their suitability for participation in the program.[80] Before the end of the year, the United States had sent six experts to China under the program, followed by sixteen more in 1943, reaching a total of thirty between 1942 and 1946.[81]

The State Department experts were sent with the explicit understanding that, unlike experts sent by other U.S. government agencies such as the Board of Economic Warfare or through lend-lease, they would be serving the Chinese government rather than the U.S. government or the Chinese military effort.[82] At the same time, the experts were to be under the supervision of the U.S. embassy, which was required to make reports on their activities.[83] In China, these experts linked up with government organs where they helped to conduct surveys, plan development programs, and train personnel. Although not all of the experts found conditions in China to be sufficiently developed to allow them to make significant contributions, and some of them did not react especially well to being in China, others worked closely with Chinese partners to develop programs that they believed would, or at least could, have a lasting impact on the region.

Not to be outdone by the Americans, the British Foreign Office also sent experts and looked to send more. Joseph Needham is perhaps the best example of such an expert, although he was sent not as an adviser attached to any Chinese government organ but rather as an employee of the British embassy, whose job was to facilitate Sino-British scientific cooperation. Needham, who was sent to China in 1942 and remained there until 1946 running the Sino-British Science Cooperation Office, spent much of his time traveling throughout unoccupied China visiting universities, research institutes, and state-owned enterprises where he observed ongoing scientific and technical activities, cataloged requests for scientific papers and laboratory materials, and gave lectures on Sino-British science cooperation as well as his own areas of expertise. With the exception of his lectures, few of Needham's activities in China strictly fell under the category of training. However, his curiosity to learn and his energetic mind, combined perhaps with a certain degree of self-certainty, led him to eagerly offer advice on all sorts of scientific and technical topics so that at some level, he found himself functioning as a sort of collaborator in a wide range of scientific activities in inland China.

One of Needham's roles as a British representative to China was to send home reports on the things he observed, some of which, on topics such as roads and oil wells, were classified, and others of which were published in such venues as the journal *Nature*. Among his reports to the Foreign Office was a January 1944 "Memorandum on Western Technical Experts in China" in which he described the State Department program in some detail. A particular matter of concern for both Needham and the Foreign Office was the question of postwar influence and whether the Americans were, through the cultural relations program, putting themselves into a better position to benefit from a postwar economic relationship with China than the British. Needham, who had met and spent time with many of the State Department experts, clearly felt that their interest was squarely focused on the development problems of the moment, rather than those of the future. However, he also felt that the Chinese were more receptive to American experts, saying that Chinese believed their motives to be altruistic, whereas they perceived the British in China to be acting out of self-interest and motivated by a desire to establish a postwar economic foothold. Not surprisingly, his Foreign Office colleagues viewed Needham's perspective on U.S. assistance to China as naive, noting in their handwritten comments attached to his letters and reports that they were not so sure that the Americans were not also in China to develop relationships that would serve U.S. economic interests after the war.[84]

Britain, also facing a shortage of funds and personnel during the war, never developed a systematic program for matching experts to government needs to rival that of the State Department's Cultural Relations Office. But the British Council, a British government organization that aimed to promote cultural and educational opportunities around the world, did at least work to build up its Sino-British Science Cooperation Office into a full office with a regular staff of scientists who would be able to serve as consultants and manage other scientists and technicians who would later be sent to address specific technical needs in outlying regions. From the council's perspective, the office would provide "a central base from which individual experts in outlying places can be directed and supervised, and organized to the best advantage." To staff the office, Dorothy Needham, Joseph Needham's wife and a recognized biochemist in her own right, as well as William Band, a physicist, were sent in 1944. As of June 1944, a biologist, Dr. Picken, was awaiting transport to Chongqing, and the British Council was working to identify an engineer, an agriculturalist, and a medical scientist to serve on the staff of the office and were starting to look for experts on wheat taxonomy and library representation to send as experts to help with specific problems. They had also made some effort to find an expert in glassblowing but had had no luck because of the shortage of such personnel in Britain at the time.[85] Other British experts did travel to and within China, but for the most part, they appear to have been there more to make reports back to Britain about conditions in China than to provide concrete technical assistance. In addition to the classified reports Needham filed, Sir Erik

Teichman sent back numerous reports on the state of roads in China's northwest, and John Blofield returned reports on the state of higher education.

The Foreign Office, however, was less enthusiastic than the British Council about the prospects of assisting China with technical experts not because they thought such expertise should not be offered but more because of the shortage of both funds and personnel in Britain to support such an endeavor and because the Chinese were already securing experts from the United States, who could be supplied in greater abundance. Nonetheless, by mid-August 1943, in response to a Chinese government request for technical experts, the British government was in the process of identifying experts in a few fields. In September, the two sides were in negotiations regarding experts in electrical power and the manufacture of alkalis and fertilizers, and Weng Wenhao, minister of economic affairs, had expressed to British consular officials a hope for advisers working on shipbuilding, dyestuffs, and plastics. As of that date, however, the British Foreign Office had still not authorized Ambassador Horace Seymour to confirm to the Chinese government that the British government would pay the salaries of the British experts.[86] By the end of the year, British officials had the impression that their offers of technical personnel had been rebuffed owing to the fact that the British wanted their experts to be helping to draft plans for postwar reconstruction rather than focusing on immediate development needs.[87] From Needham's perspective, though, the Chinese were concerned that British experts would be too closely allied with commercial interests. Although the Foreign Office found Needham's analysis to be problematic—A. Scott observed that it was "typical of Dr. Needham that he should build up his thesis on a foundation which is by no means secure"[88]—Needham, who spent a good deal more time than the Foreign Office men talking with both Chinese government leaders like Weng Wenhao and Chinese scientists and technicians, appears to have had a much better grasp on the legacy of British imperialism on the collective psyche of the Chinese government than did his Foreign Office colleagues. Foreign Office diplomats seem to have been most strongly motivated in these efforts by a desire not to cede influence over China's scientific development to the United States, and if their Chinese counterparts perceived such motives in their actions, it would be little wonder that they might regard British offers with some skepticism.[89] In the end, the British did not supply anywhere near the number of foreign experts to assist with training that the United States did through the State Department cultural relations program.

This debate between the Foreign Office and Needham highlights an important dimension of all soft-power programs such as those undertaken by the British Council and the State Department. As Wilma Fairbank noted in the introduction to her 1976 study of the cultural relations program, "Intercultural relations used as a tool of government policy are subject to manipulation through an entire range of motives dependent upon the good or evil intentions of the power concerned."[90] Although both Britain and the United States were surely motivated at least in part by the desire to aid an allied power to better defend itself against a common enemy, both governments and the people they sent as their representatives also had other interests. Needham may have been correct in his assessment of how his Chinese interlocutors viewed the British and American offers of assistance, but that does not mean that the Chinese with whom he spoke were correct in thinking (if they did) that the United States had purer motives than Britain for engaging in scientific and technical exchange. Moreover, the Chinese government's motives were also complex and driven largely by the need to develop the entire economy of the regions they controlled. Those complexities were not always clearly understood by their British and American counterparts.

U.S. CULTURAL RELATIONS PROGRAM EXPERTS

Among the cultural relations program experts, three stand out as having been particularly successful in developing training programs that were targeted at very specific development goals: Theodore Dykstra, Walter C. Lowdermilk, and Paul Eaton. Dykstra, a potato expert with the U.S. Department of Agriculture, spent nearly two years, from December 1942 until August 1944, in China working with the NARB to develop a seed potato program that aimed to improve and expand potato output throughout western China. Part of the work of creating a sustainable potato program, one that would continue after his departure from China, was the training of people to undertake such a project. To this end, Dykstra engaged in different kinds of training activities with both scientific professionals and farmers.

For technical professionals, Dykstra organized two kinds of formal training activities. In April 1944, Dykstra ran a two-week training course "on all phases of production" in Chengdu. The program was approved by the minister of agriculture with the expectation that representatives from the different provincial agricultural experimentation bureaus would attend. Dykstra spent months preparing his own lectures and working with Chinese colleagues to organize the program. His plan was to have colleagues make reports on their work, demonstrations of techniques, trips to nearby potato-growing areas, and participate in roundtable discussions "for a free exchange of ideas." His expectation was that the conference would provide more than just an opportunity for him to pass along technical information to colleagues, however. He also wanted it to provide an opportunity to work with the potato group to develop a comprehensive potato program.[91]

Unfortunately, as Dykstra noted in his diary, "On account of poor transportation only Mr. Chiang from Wukong representing the northwest was present as the only outsider. Mr. Lee, from Wening was unable to come on account of poor transportation. It was also difficult to get workers on account of planting time." Most of the attendees were agricultural technicians already in residence in Chengdu, people with whom Dykstra already had regular contact. In spite of the fact that not all of the target trainees were able to attend, the conference went on, and Dykstra reported that "lectures were given in the morning and were attended by the regular members of the potato group" as well as "by Mr. Wang of the vegetable department of Nanking University, and by Mr. Hsi of the Horticulture Department of West China Union University. Both of these men are planning on joining the project sometime in the future. In addition to this some of the technical workers of the Provincial Station, and some students of Szechwan National University also attended."[92] The training course did not pan out precisely as Dykstra and the minister of education had envisioned it, but it did go forward, and at the very least, it provided an opportunity for Dykstra to extend training to a number of attendees who were not already part of the potato group with whom he worked regularly.

Also, during the winter of 1943 and spring of 1944, Dykstra spent a number of weeks in Beibei, Chengdu, and Chongqing giving lectures at area universities. In Chongqing, he presented a series of two-hour lectures twice weekly over a period of several weeks at National Central University where he reported that his audience on the first day consisted of fifty-two students, nine of whom were female, who were all junior and senior horticulture and agronomy majors. In addition to the students, according to Dykstra, the lectures were also attended by half a dozen staff members. At the same time that Dykstra was giving his lectures, three other State Department technical experts were also lecturing on their own areas of specialization. Paul Eaton, a mechanical engineer, George Cressey, a geologist, and Fred O. McMillan, an electrical engineer, all gave their own lectures at National Central University. The four experts were all picked up together and transported to the university and then returned to the Sino-American Cultural Relations Institute where they were staying.[93] A few weeks later, Dykstra gave another set of eight lectures, perhaps the same ones, at Jinling University, which was then located in Chengdu. Dykstra's goal with all of these activities was to "help to train some personnel to solve potato disease problems," an endeavor that he took very seriously. Intent on offering the highest quality lectures he could, he moved for a time from his base of operations in Beibei to Chongqing where he could use a microfilm reader at Zhoujin Middle School to prepare his lectures for both his training course and his university lecture series.[94]

It is, of course, impossible to know just how impactful these lectures were. Dykstra appears to have felt that his lectures made a difference. According to Needham, however, others among the cultural relations program experts were not convinced that their lecturing had much of an effect. Needham wrote in his 1944 memo on technical experts that Fred McMillan, for example, "was less certain of the value of his lectures there than of the value of his visits to nascent industrial undertakings, where he was able to advise Chinese engineers as a consultant."[95] Visiting too many plants, however, might potentially dilute the efficacy of any training undertaken by the technical experts, as pointed out by another of Needham's confidants, S. A. Trone, a Russian American engineer who served as adviser to T. V. Soong. Trone told Needham that he felt that the U.S. engineers would do "much better if they were to concentrate on one particular plant, and stay there as long as possible, actually training the Chinese engineers and technical foremen by working with them." As Needham noted, some of the cultural relations program technical experts, particularly the three who had been seconded to the Yumen oil field in Gansu, were doing just that.[96]

Like the oil drillers stationed at Yumen, Dykstra also spent a good deal of time working directly alongside his Chinese colleagues and much of the work he did with them could also be described as training both for his colleagues and for the farmers they encountered. Dykstra traveled with his Chinese colleagues to inspect potato fields all over the region, looking for disease-free fields and finding few. He worked with them to plant tubers he had brought from the United States in fields located across western China, and together, they evaluated the degree to which those strains were well adapted to the varied environments of the region. His primary collaborator was Dr. C. C. Kwan, a Cornell PhD who had been assigned to lead China's potato project but who spent much of Dykstra's early time in China trying to extract himself from his prior work in Guiyang.[97] Dykstra praised Kwan as "a man of very good judgment, and a valuable companion" on expeditions to survey potato fields, noting that "we agree in all subjects pertaining to the development of a national potato program in China, 100%."[98] He says comparatively little about his other collaborators. We know, for example, that he traveled on his expedition to the northwest with Mr. Chang and Mr. Yang of the Gansu Provincial Experiment Station, but he tells us little about them (including their given names). Nonetheless, Dykstra's diary entries make it clear that these technicians were active partners in the enterprise (see fig. 2.4), sometimes splitting off from Dykstra so as to allow the party to survey more fields.[99] He also worked with technicians such as a Mr. Chiang, who helped him plant potatoes in experimental plots near Chengdu.[100] Most of these men already had some degree of expertise when they began working with Dykstra, but Dykstra's specific expertise in potatoes allowed him to offer them advice and training that he believed could help them to develop a workable national potato program.

The farmers they encountered during their outings were also potential trainees, and the NARB, with which Dykstra was working, certainly understood one dimension of the national potato plan to be training for farmers on seed potatoes and cultivation methods.[101] In fact, those farmers were critical to the entire project. When Dykstra, Chang, and Yang went out to do fieldwork, their primary goal was to identify existing potato farms that were cultivating plants without disease and to work with

FIG. 2.4. Dykstra's colleagues and a farmer at a potato farm. Courtesy of Harvard-Yenching Library of Harvard College Library, Harvard University.

those farmers to preserve their potatoes for use as seed potatoes to be planted elsewhere during the next season. A secondary goal was to educate the potato farmers with diseased plants on how to prevent disease. In both cases, they were attempting to enlist the farmers as active partners in a larger-scale plan to improve the quality and quantity of potatoes being cultivated in China. Chang and Yang played a crucial role in this process, not only for their expertise as potato scientists but also because they were the ones undertaking the complex negotiations with farmers who were reluctant to see their fields culled of diseased potatoes or who needed to be helped to understand how they might contribute to the production of a national seed bank (see fig. 2.5).[102]

Dykstra's own descriptions of his activities in China, written in detail in letters to his wife, almost universally suggest a one-way transmission of knowledge to China. He aimed to set up a national potato program and to train people to run it. He also aimed to "write a book covering all phases of potato production for China" that could be "used as a textbook in universities and as a handbook for technicians." Having done these things, he told his wife, "I can go home with the feeling that with the whole hearted cooperation of the Chinese Government, the foundation of a sound

FIG. 2.5. Farmer Yang in his potato field. Dykstra and his colleagues had protracted negotiations with farmer Yang over the importance of culling sick plants from a field that would be producing seed potatoes for the potato project. Courtesy of Harvard-Yenching Library of Harvard College Library, Harvard University.

potato program in China has been laid."[103] Much though these words suggest that Dykstra saw himself entirely in the role of guide and instructor, in fact, his letters and diary reveal that he was also constantly learning while in China. In China, he encountered numerous new varieties of potatoes and learned from his colleagues and from farmers a great deal about regional cultivation and uses of the tubers. This knowledge shaped his ideas about what could constitute a viable potato program for the nation.

Another cultural relations program expert was Walter C. Lowdermilk, an American soil erosion specialist who had spent time in China working on soil erosion as a younger man and who returned in the fall of 1942 at the request of the Chinese government to help the Chinese fight erosion and grow more food. He returned to the United States in early 1944. While in China, Lowdermilk undertook an extensive survey of land use in northwest China. As Lowdermilk wrote in his preliminary report on his activities, "The object of this survey was twofold: First, to give technicians and officials of the Chinese Government the benefit of our experience in the United States in finding the facts about erosion, in developing, organizing and carrying forward a nation-wide movement in soil and water conservation, in erosion and silt control, and in increasing by such measures crop production on cultivated, grazing and forest lands. Second, to designate a number of suitable areas for demonstration projects, to work out practical measures and to put them on the land for farmers and technical men to see."[104] Training, broadly understood, was therefore very much part of Lowdermilk's remit. Like Dykstra, much of the training occurred as he worked side by side with Chinese colleagues and farmers. Also like Dykstra, even though Lowdermilk's words suggest that he, too, expected to be engaged in an entirely top-down instructional enterprise, in fact, his proposals to combat soil erosion in China were also heavily influenced by the practices of Chinese farmers he observed. Nonetheless, he clearly understood the people with whom he worked to be his staff (a word that Dykstra did not employ to describe the Chinese technicians with whom he worked).

To conduct his survey, Lowdermilk requested a field staff of people with expertise in the varied fields of soil science, agronomy, agricultural engineering, forestry, agricultural economics, and grassland management. Eight specialists were seconded to his group, two of whom were from the NARB, one of whom was an engineer in the MOAF, three of whom worked on soil and water conservation under the MOAF, one of whom worked for the National Conservancy Board, and one of whom worked for the Gansu Agricultural Improvement Bureau. All of these men had at least a BS, and five of the eight had master's degrees. Six of them had studied abroad, and most of them spoke English fluently, making Lowdermilk's job easier.[105] Lowdermilk spent three months in Chengdu training his "devoted and diligent" staff to conduct a land-use survey and in American principles of soil and water conservation before heading off into Shaanxi and Gansu. He also "gave special emphasis to adaptation of findings in America to conditions in China," a point that he believed to be particularly important. American techniques and methods, Lowdermilk felt, could form a useful foundation but should not be applied without consideration of how local conditions might require adaptation.[106] To do this, his team needed to get out in the field, but in Lowdermilk's telling, not all of his staff were initially enthusiastic about that prospect. In particular, he believed that Y. H. Djang, the group's agricultural engineer, just "wanted to be a scholar." As Lowdermilk observed, "He said to me, 'You get the facts and I'll make the design here in the office.' I said, 'But

I want you to get out in the field and apply your design to the local situation.'" Lowdermilk certainly believed that his insistence had a transformative effect on Djang. "He preferred desk work then," Lowdermilk said rather boastfully, "but now he works in the field."[107] Whether Djang was truly resistant to fieldwork or not is impossible to know. We do know, though, that after the war, he continued to be active in the field and went on to work for the Joint Commission on Rural Reconstruction in Taiwan and later assisted on hydrology projects in Africa for the United Nations Food and Agricultural Organization.[108]

One of Lowdermilk's big points of emphasis with his Chinese staff was that they needed to listen to farmers. "In working with my staff, I set up an objective, to evaluate farmer practice. I said, 'These farmers have been farming this land for several thousand years, and they've built up traditional knowledge based on trial and error from times past. They have learned what to do but their explanations do not sound very reasonable to a man trained in modern sciences, such as physics, chemistry and biology.' The tendency of the scientist had been to say, 'Well, these farmers don't know the facts, they're old-fashioned, so we'll discard this past knowledge and start anew.'" However, Lowdermilk's perspective was that "all we need now is to collaborate. We'll take these measures that have worked out, and we'll use our scientific procedures—surveying, exact levels that we want, and so on."[109] We can see that Lowdermilk had considerable enthusiasm for learning from the farmers, but his words also convey how he felt about the attitudes of his staff. Whether it was true or not, Lowdermilk understood his staff to be so wedded to modern scientific techniques that he believed they would not be willing or able to consider that such techniques might best serve China if adapted to local knowledge. He also saw at least one of them as being too bookish to be willing to get out in the field. Since Lowdermilk is our only source on these matters, it is difficult to know if he was simply applying that old Western trope about Chinese being unwilling to get their hands dirty to his team, or if the men he was working with really were as limited as he thought them to be. As we shall see in chapters 4 and 5, however, plenty of Chinese scientists and technicians well understood the importance of both adaptation and learning from local experts and did not require the kind of pushing that Lowdermilk and other American observers seemed to feel was necessary.

To a certain extent, Dykstra and Lowdermilk exhibited the characteristics of the nineteenth-century Western naturalists described by Fa-ti Fan. Fan's naturalists understood themselves to have expertise that the locals with whom they worked lacked. They "valued honesty, facts, and truth" and were, thus, in their own minds, "the opposite of the Chinaman."[110] Dykstra and Lowdermilk may have brought to China a set of inaccurate assumptions about Chinese scientists and technicians and a sense of their own innate superiority, but they did recognize that their colleagues were trained scientists and technicians. Nevertheless, like Fan's naturalists, they often failed to acknowledge the extent to which they depended on Chinese scientists, technicians, and farmers as they undertook their projects.

Lowdermilk's team surveyed land use and the topographical features of the land, noting such things as the gradient of the land, the way crops were being planted, and the type of runoff and erosion the land was experiencing. They also took note of existing planting strategies (such as pebble mulching, which was being used near Lanzhou). Having collected such information, they established experimental plots in locations that had been predetermined by the MOAF, where they set up demonstration projects. For example, on a sloping terrain in southern Gansu, they planted experimental strips of various crops such as alfalfa, rye, and sweet clover to determine which crop(s) might absorb the most rain. They also tried narrow strip cropping on the contour with a contour channel to prevent runoff on slopes with a grade of up to 24 percent. These experimental plots revealed that the methods Lowdermilk's team employed were extremely effective in preserving moisture and preventing runoff. Lowdermilk viewed these demonstration sites as providing evidence "that farmers may see for themselves, and to show technicians what can be done under practical farming conditions." Just how widely his sites were toured by either farmers or technicians, however, is unclear. His expectation was that eventually the proven strategies could be disseminated through village-level farmers' associations that would facilitate the large-scale spread of the methods.[111]

From a technical training perspective, not only did Lowdermilk train a small number of educated young scientists who were already engaged in either teaching or bureaucratic careers, but he and his team went on to train farmers through their demonstration projects, which served simultaneously as experiments to determine what strategies would work best and demonstrations of techniques that they hoped to encourage farmers to start using. It is unclear whether any of the farmers Lowdermilk's team worked with continued to use these techniques, but it appears that at least some of the young technicians on Lowdermilk's team continued to do hands-on work in China, Taiwan, and even Africa, though others worked their way up the bureaucratic ladder. Ren Chen Tung, who worked with the team but was not among the eight experts originally seconded to it, was the one about whom Lowdermilk spoke most effusively in his oral history. According to Lowdermilk, he "developed a successful model community in the upper Yellow River near the big loop, with schools and full employment. He was a farmer's son and became a modern farmer and trained others to be expert farmers."[112]

Unlike Dykstra and Lowdermilk, Paul Eaton, head of the Lafayette College mechanical engineering department, was sent to China specifically to work as an industrial trainer. Prior to his departure for China, in fact, he worked to help the NRC locate possible placements for as-yet-unidentified Chinese trainees by writing to sixty U.S. factories to inquire if they would be willing to offer practical training to Chinese students and engineers.[113] Once in China, in addition to conducting site visits to "several hundred plants located in Szechwan, Kweichow, Kwangsi and Hunan Provinces" to observe apprentice training strategies and labor organization, Eaton also taught a course for industrial trainers.[114] Eaton offered the "Job Instructor Training" course to several hundred Chinese engineers who took the twenty-hour evening course lasting ten sessions in groups no larger than thirty in Chongqing as well as Yishan in Guizhou and Hengyang in Hunan. The goal of the course was to train the trainers in best practices in industrial training. The course focused on educational psychology, the principles of job training as laid out by the Western Electric Company training manual, and effective foremanship. Eaton's students were working in a number of different industries, and he generally found them to be "anxious to increase production in China" and enthusiastic to participate in the course. Following his return to the United States, Eaton learned that "the men whom I taught are now teaching others who in turn will train foremen," and he observed that "this news is encouraging for it indicates a sustained spirit of enterprise which will keep the program moving forward."[115]

Sustained training that would, over time, have a transformative effect on Chinese industrial management practices was in fact of great interest to Eaton. He recognized that American labor practices could not be transferred wholesale to China, where, in his observation, laborers answered most immediately to "gang leaders" who had neither the education nor the social stature of American foremen, and who were thus paid less, given lesser managerial roles, and had virtually no opportunity to advance into higher management. The gang leader, he argued, "must assume a higher place in industry and in society if Chinese industry is to reach its full potential development." On the other hand, Eaton felt that many of the factories he had visited had developed strong apprenticeship programs for middle school graduates that would benefit most from modernization of classrooms and workshops and greater abundance of modern machine tools, which were all in short supply in the factories in western China. As for new training programs, he felt that China was ready for "an industrial training course along lines paralleling the splendid work of the War Manpower Commission in the United States" but with modifications to suit "the different industrial and social background of the two nations."[116]

Although Eaton was not able to facilitate the creation of this kind of industrial training course in China, he continued after his return home to work in various ways to facilitate training of Chinese in the United States. In May 1945, L. F. Chen, of the NRC New York office, approached Eaton to request his assistance with a large group of NRC trainees who would be coming to the United States later that summer. Chen was responsible for finding placements for the trainees in the industry and mining division, which was the largest single group of trainees, numbering in the hundreds. Anticipating that it might be hard to find placements for all of them to start immediately after their arrival, Chen wanted to find a way to help them utilize their time in the United States effectively right from the start. For this reason, he wrote to Eaton asking for his help in setting up some activities for the trainees. Chen's initial request was modest. He was simply looking for colleges where he could situate the trainees temporarily and where they might be able to sit in on some classes and use the library. But Eaton took a much larger role than Chen initially anticipated, helping to organize a formal training program and traveling to Washington, DC during the summer of 1945 to teach in it himself.[117] In this manner, Eaton did realize his ambitions for a Chinese industrial training program; it just did not take place in China.

As we can see from these examples, these technical experts provided training services while they were in China, and at least some of them continued to work with their Chinese colleagues even after they departed or to participate in further training activities for Chinese who came to study in the United States. Following his return to the United States, Eaton kept in touch with the Chinese Institute of Engineers, the NRC, and the Universal Trading Corporation, all of which were Chinese organizations in the United States. He spoke at Chinese Institute of Engineers, America Section meetings, helped with placement of Chinese trainees in the United States, and orchestrated a drive to collect technical literature to send to China. And as we have seen, L. F. Chen also sought his advice on how to set up a preplacement training program for NRC and lend-lease trainees who would be coming to the United States starting in the summer of 1945.[118] Lowdermilk helped Feng Chao Lin, one of his Chinese colleagues, to come to the United States for a PhD and remained in touch with a number of his Chinese staff for many years after the war.[119] Other cultural relations program experts also continued to maintain relationships with individual Chinese, including representatives of the Nationalist government, long after their return to the United States.

Conclusion

A variety of training activities that aimed at expanding the number and deepening the expertise of scientific and technical personnel took place in China during the war. These activities were targeted both at addressing the problems of the moment and putting China in a position to be able to modernize its postwar economy rapidly. The Ministry of Education attempted to transform the educational system in ways that would produce more highly educated scientists and technicians. Other programs were directed at other audiences including civil servants and industrial technicians on the one hand and lay practitioners such as farmers and factory laborers on the other hand. These programs were devised by a variety of actors, including Chinese educators, researchers, and bureaucrats as well as foreign social activists and technical experts. A primary goal of such training programs was typically to train new technicians, agriculturalists, laborers, and managers or to help existing ones to develop skills that they needed to address immediate needs, but nearly all of the leaders of such programs were also thinking about the long term.

Foreign entities involved in the development of such programs engaged in these activities with a variety of motives. The foreigners leading the CIC movement and setting up the Bailie Schools, for example, were interested in transforming Chinese society and the Chinese economic structure by leading a bottom-up educational movement for the masses. In their vision, both Bailie School students and CIC trainees would develop, through active learning, the knowledge and skills necessary to permit them to become trainers themselves, thus giving the trainees and students ownership over an ongoing process of social and economic reconstruction. The British Foreign Office, hampered by Britain's own wartime constraints, seemed more interested in gathering intelligence than in initiating programs to support Chinese development. However, the British Council, also government funded, did set up an office, staffed by Needham and several others, to foster scientific and technical cooperation by sending British scientific publications, giving Chinese researchers international venues for the publication of their own work, and providing small amounts of support in other ways.

The U.S. government, through the cultural relations program, tried employing soft-power strategies with China that it had been using with Latin America since 1939. The timing and circumstances of the China program, however, oriented it away from cultural exchange and toward scientific and technical engagement, which better aligned with White House priorities as well as Chinese needs. Within the State Department, the rationale for the whole Division of Cultural Relations was the extension of American soft power and cultural values. Even with the China program's reorientation toward science and technology, these elements were always there. By extending this nonmilitary aid to China, the State Department concretely demonstrated American friendship and built human relationships between Chinese and American government officials, scientists, and technicians. These relationships, particularly because the resources that supported them were given during China's hour of need, had considerable potential to bring China more squarely into an

American political and economic orbit—much as staff at the British Foreign Office feared.

However, these programs all had their limitations. Relatively few foreign experts could be brought to China and the extent of their reach was limited. Extensive efforts to revamp Chinese education were hindered by limited human and other resources on the one hand and a rapid expansion of the number of programs on the other hand, leading to a serious dilution of instructional capacity. As a result, the push to produce more middle school and university graduates in specific fields came into tension with the need to produce graduates who had the actual knowledge required to meet China's industrial and agricultural personnel needs. Given these problems, government planners increasingly looked to training abroad as offering the best solution for increasing the number of higher-level technicians that were needed to address immediate problems and that would be required in even greater numbers to implement postwar reconstruction plans. Chapter 3 will explore this topic.

CHAPTER 3

Developing Human Resources

Education and Training in the United States

As we have seen in chapter 2, human resource development was a major preoccupation for both the government and other groups interested in promoting development in wartime China. Government institutions such as the National Resources Commission (NRC) and the Ministry of Agriculture and Forestry (MOAF) had an immediate need for skilled industrial and agricultural technicians who could help get the wartime economy up to speed. Entities such as the Central Planning Bureau (CPB) and the Ministry of Education also engaged in longer range human resource development planning, anticipating postwar needs. These institutions created a variety of different types of training programs in China, sometimes with the assistance of foreign technical advisers. The higher-level training that would be required to truly modernize both industry and agriculture, however, was hard to produce in wartime unoccupied China because laboratory equipment and industrial machinery tended to be outdated or of poor quality, literature on modern technologies was scarce, and good models of modern technologies in action were even rarer. For these reasons, especially as the war went on, the Nationalist government grew increasingly inclined to create training opportunities abroad, even though they ran the risk that trainees would simply not return or that the cost of training would not be recouped through follow-on knowledge transfer after the trainees returned.

The government partnered in these efforts with Chinese and Chinese Americans in the United States as well as U.S. government and nonprofit institutions and U.S. businesses and academic institutions. To manage these efforts, the Nationalist government set up both public and private semipermanent entities in the United States and encouraged the revitalization and creation of some academically oriented societies. All of these groups were at least somewhat motivated by a desire to help China address its immediate wartime needs, but most of them were also acting in their own interests, interests that were frequently more focused on postwar opportunity than on the war.

Study Abroad in the United States

Although more costly and in many respects riskier than study or training in China, training or study abroad offered the best way to obtain the in-depth knowledge and skill development that government organs needed to carry out their plans. Students had the opportunity to enroll in courses that were not offered in China and to work in better laboratory facilities than they would have found at home. Trainees spent time in factories, agricultural extension facilities, and government agencies learning about scientific techniques and organizational strategies that they could apply at home. Through these experiences, both students and trainees were exposed to knowledge and applications of knowledge that were uncommon in China and became conduits through which that knowledge could be transferred to China.

During the war, the Nationalist government facilitated and regulated study and training abroad for both students and professionals. Concerned that Chinese students heading abroad in the early part of the war were inadequately prepared both financially and linguistically and wishing also to guide the direction of study abroad in much the same way as it was attempting to guide curricular development at home, the Ministry of Education sought to gain control over the process by determining which students could go abroad.[1] Starting in 1942, therefore, new Ministry of Education rules stated that eligibility for study abroad would be, for the most part, determined by government exams in English as well as disciplinary content. In addition, students were obliged to demonstrate sufficient financial support to maintain themselves for the duration of their courses of study. Restrictions were also put in place to limit study abroad to graduate students, and over time, further restrictions were made on content so that by the end of the war, very few students in social science and humanities disciplines were permitted to go abroad and the vast majority of study abroad students were doing graduate work in scientific and technical fields or management. Students preparing to travel were also required to undertake a predeparture training program overseen by the Central Training Institute.[2]

U.S. observers believed that the Nationalist government, in addition to implementing new rules governing who could study abroad, was also attempting to control the political activities of students while they were in the United States. Rumors that the Kuomintang (KMT) was putting political pressure on students in the United States and that it intended to appoint a superintendent of students to control students' thought reached the State Department in April 1944, which through a conversation between Ambassador Clarence Gauss and T. V. Soong, made it clear that from the U.S. perspective, such political oversight would not be tolerated by the U.S. government.[3] In response to the U.S. pressure, Chiang Kaishek somewhat impractically "issued orders stopping all students and technicians from going to the United States."[4] The Chinese government ultimately altered the regulations to eliminate the superintendent of students position, and by September 20, the ban on students going abroad had been lifted.[5]

Whether or not the superintendent of students position was intended to perform the political function the State Department feared, the fact that its creation generated this sort of response underscores some of the tensions in the Sino-American wartime relationship and highlights some of the challenges the Nationalist government was facing. From the position of a state that was seeking to exert greater control over education, the freedoms accorded to students in the United States and Europe seemed problematic. The Nationalist government wanted to ensure that the students it supported were being trained in the fields that the government most urgently needed, and oversight by a supervisor of education might have helped keep students abroad on track. Of course, the Nationalists had also already demonstrated through their efforts to control their own institutions of higher learning that they did not want students (or, for that matter, faculty) to get too politically active. Americans in China frequently evinced concern about crackdowns on political activities in Chinese in-

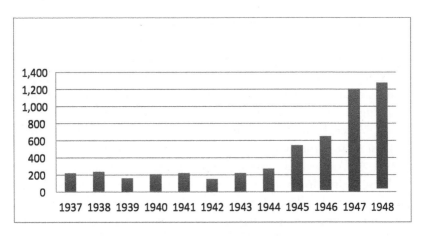

FIG. 3.1. Number of students in the United States by year of entry. Source: Mei, *A Survey of Chinese Students in American Universities and Colleges in the Past One Hundred Years*, 25, 27. Mei's data are based on a survey that the China Institute sent to 704 U.S. colleges and universities. They received responses from 560 institutions.

stitutions of higher learning. Thus, their position on the superintendent of students is not surprising.

During the war, most Chinese students pursuing degrees outside of China did so in the United States, though some still went to Britain. Most of the other common prewar study abroad venues were no longer appropriate destinations during the war. As one close observer noted in 1943, because of wartime obstacles to studying in Europe and even Canada, "the United States must be considered as, and has become, the only place where training for technical personnel for wartime China and post-war reconstruction in China can be carried out to any practical extent."[6] On the whole, the number of Chinese students traveling to the United States to study grew steadily over the course of the war (see fig. 3.1). There were dips in both 1939, owing, almost certainly, to the incredible disruption to Chinese higher education that had taken place in the previous year, and 1942, owing to government interference in the process. Over the course of the last several years of the war, and heading into the early postwar period, the number of students traveling to the United States grew quite rapidly.

Many students were motivated to study abroad by poor living and inadequate learning environments at home. As one student told Edwin Kwoh in an interview for his 1946 EDD dissertation on Chinese students in the United States, "In China food is scarce and expensive during the war. It is no fun to study on an empty stomach. So, if there is any chance of coming to America as a student, to be supported by the government, everyone is eager to get it."[7] Not only did students have to deal with poor living conditions in China, but remaining a student in China also meant studying in universities with subpar learning environments. As Kwoh noted, "After the war started in 1937, the Chinese colleges and universities lost most of their equipment and library books on their trek to West China. An American newspaper correspondent visiting a refugee college in West China discovered that, on an average, twenty students had to use one book. Intellectual starvation at home due to the lack of books, laboratory equipment, etc., has driven hundreds of China's young people to America to pursue their education in well-equipped foreign institutions. This motive for attaining intellectual training is eminent among the present Chinese students in America."[8]

Desire to serve the nation, particularly in a time of war but also at a moment in which many were looking ahead to the possibilities and opportunities that would arise during the anticipated postwar period of national reconstruction, was certainly another strong motive for study abroad. Many students sought to contribute to specific development projects that would strengthen China. As Yaozi Li 李耀滋 said of himself and other Chinese students who received scholarships to study aerospace engineering at the Massachusetts Institute of Technology (MIT) during the war, "All of us received our PhD degrees rapidly and went back to China during WWII to join with the rest of our classmates in the effort to help China develop her infant aircraft industry and technology."[9] Li and many of his colleagues in aerospace engineering did in fact return to build China's airplane program during the war, many of them truncating their studies in order to do so. Students in other fields were also recruited by NRC representatives in the United States into training programs in U.S. industries or to return to China. The National Bureau of Industrial Research (NBIR), for example, actively identified possible candidates and wrote to them, asking that they return to China and, as one respondent to such a letter wrote, "take part in the reconstruction and industrialization of our great country."[10] Some students took them up on these proposals.

Although it is surely the case that many more students were electing on their own to study applied sciences because their career options in such fields seemed brightest, it was also unquestionably the case that the Nationalist government was looking to funnel students into such fields. As Chih Meng, director of the China Institute in America, observed in 1943, "During the last ten years, the trend has been to send only graduate students from China to do post-graduate work in America. The duration of their stay is considerably shortened. These graduate students have been selected and instructed to come here to do highly specialized research in the shortest period of time possible. Meanwhile the subjects of study have also become more selective. In other words, there has grown in China a realization that certain fields of study or research can best be obtained in this country, while certain other fields are perhaps of no immediate importance. The subjects selected are principally in the field of applied science and technology."[11] The "important" fields were those of immediate relevance to China's reconstruction efforts, and students in those fields were expected not to linger in their studies but to return as quickly as possible to China where they could begin to apply their new knowledge. "In normal times," Meng observed, "in a democracy such as the United States or China, it is probably helpful to pursue more or less a laissez faire policy toward personnel training. But in war time a post-war reconstruction maximum efficiency can only be achieved when individual talents and desires are coordinated and utilized for national objectives and objectives of the United Nations. In other words, a fairly complete program of co-operation and co-ordination must be worked out between national planning and its personnel requirement."[12] Meng, who was not himself affiliated with the KMT, nonetheless saw that centralized planning for personnel development should, at least in times of emergency, be the role of the state.[13]

One way in which Meng, himself, facilitated such planning was by arranging support for Chinese students in the United States during the war. Students already in the United States when the war broke out often ran out of funds to support their continued study and were unable to secure passage back to China even when they had managed to complete their degrees. In 1938, seeing the need for emergency aid to these students, Meng, in his role as director of the China Institute in America, worked through Eleanor Roosevelt and Lauchlin Currie to establish and run a State Department emergency aid fund through the China Institute. Meng formed a three-person scholarship committee, the State Department appointed its own counterpart committee, and the two groups working together made 1,666 scholarship grants to students and 474 grants to support on-the-job training, typically for recent graduates engaged in practical training in technical fields.[14] Grants were intended to "carry out the wartime planning of personnel training." Fellowships for graduate students who were either engaged in research or interested in pursuing training in industry were awarded on the basis of three criteria: "1. The immediate importance of the student's field of specialization to China; 2. The student's ability; 3. The student's financial need."[15] In addition, Meng was not averse to encouraging students to switch directions after graduation so as to develop expertise in a high-demand field that would better serve China's reconstruction needs. Seeing that travel on the newly constructed Burma Road was impeded by road and truck maintenance problems, Meng also actively "recruited Chinese student volunteers," some of whom were Chinese American undergraduates and high school students, "to train as mechanics for maintenance of the Burma Road and of vehicles carrying supplies to wartime China."[16]

Many scholarship recipients, though undoubtedly happy to find funding to support completion of their studies, were also enthusiastic in their desire to serve China. As one such student wrote in a letter to the State Department, "My BS in nursing education should allow me to teach nursing in China. I hope to teach as many as possible: we do not have enough nurses. I have been here four and a half years and I am longing to go back."[17]

When the first $500,000 from the State Department ran out, Meng appealed to Currie for more, and Currie suggested that he should submit a similar appeal to the Nationalist government, hoping that they would share the burden. Meng drafted a memo to T. V. Soong that outlined the need for aid for Chinese students and received a positive response within a week. Owing to the fact that H. H. Kung and others had questions about Meng's political sympathies, however, it took somewhat longer for the arrangement to come to fruition, which it finally did with the creation of the Committee on Wartime Planning for Chinese Students in the United States.[18]

Scholarships were just one mechanism that Meng and the Committee on Wartime Planning utilized to stimulate students to contribute to China's postwar reconstruction. Meng and the committee actively helped to construct a series of national reconstruction forums (學術建國討論會)

at universities across the United States in 1942 that brought Chinese students together to discuss national reconstruction and to work out how best to put their newfound knowledge and skills to work for the nation even as they themselves remained abroad. As the constitution of the University of Michigan's branch observed, "The purpose of this organization shall be to promote discussion of existing and future problems of reconstruction in China and to foster research into such problems. The immediate undertaking of the organization shall be to render effective aid to intellectual leadership in China by the preparation and dispatch of up-to-date material to China. The larger undertaking of the organization shall be to cooperate with other groups of similar nature in China and overseas in taking up similar projects."[19] Although the group aimed to assist in the longer-term national reconstruction process, its most immediate goals were to compile and transmit new knowledge that could be useful in China's wartime circumstances. To facilitate the transmission of this "up-to-date material" to China, the group would work closely with the China Institute in America, which would prepare and send the material on behalf of the Michigan group.[20] Care was taken to recruit members who would make meaningful contributions. Like the China Institute itself, the group was explicitly nonpartisan and did not take political affiliations into account when selecting new members. Members did, however, have to have "i. Active interest and enthusiasm for future reconstruction work in China. ii. Constructiveness and initiative in spirit and ideas, and willingness to contribute to the group. iii. Good scholastic and general 'intellectual' ability and outlook, including breadth of vision to accept wide scope of problems. iv. Cooperative spirit and real teamwork qualities, including professional tolerance. v. Willingness to sacrifice a reasonable amount of time and energy towards the work of the group."[21] Similar groups were established throughout the summer and fall of 1942 at universities and colleges across the United States, all with comparable goals and structures. By August 1943, thirty-three chapters had been established.[22] Chapters held meetings at which "Chinese and American specialists and students were invited to present projects for open discussion."[23] According to Ting Ni, more than 1,700 people participated in these events over a five-year period (1942–1947).[24]

The China Institute was well suited to both play its role as coordinator of scholarship support for students and lead the national reconstruction forum movement because it was, according to Meng, "a service organization for all Chinese students in the United States."[25] From Meng's perspective, Chinese students in the United States needed an entity to bring them out of their shells and organize them. As he noted in 1942, "When the number of post-graduate students increased during 1930–42, the tendency [of students] had been toward isolation. That is, each student devoted almost his entire time to his own study or research. . . . The changed composition of the whole Chinese student body [from predominantly undergraduate to mostly graduate students] created a real need [for an organization to help students stay connected]." To meet that need, the China Institute coordinated the establishment and operation of the reconstruction forums, sending representatives to all of the institutions with large Chinese student populations at least twice a year and keeping in touch "with student centers, and with individual students, concerning their study, research, practical training, and other problems, programs and projects."[26] Meng, who had directed the institute since 1930 and who had in the 1930s conducted surveys of Chinese students abroad as well as returned students in China, had a comprehensive understanding of the number of students in the United States, the kinds of activities they were engaged in, and the possibilities for their employment upon their return to China. The China Institute collected "personal, academic and technical data of all Chinese students in the United States," making the complete data available only to "governmental and other public offices for the purposes of awarding scholarships and vocational placements."[27] These data, along with the fact that Meng was actually in the United States, positioned the China Institute to play a coordinating role that the Nationalist government itself could not have. Meng not only set up structures through which students could interact with each other and, via the institute, with their compatriots at home, but he also dispensed practical advice, urging students to think about how they "fit into the program of wartime and post-war reconstruction in China" and providing them with lists of specific fields in the areas of science, technology, and management that had insufficient trained personnel.[28]

Working through the Committee on Wartime Planning, Meng also established the *National Reconstruction Journal*, which provided a channel for Chinese students, academics, and technical experts in the United States to communicate their practical research findings and observations to other members of the group as well as to scholars and technicians back home in China. In this respect, the journal was somewhat similar to *Kexue zazhi* 科學雜誌, which had been started by the Science Society of China at Cornell University in 1914. Meng, however, viewed his journal "as an instrument for the coordination of higher learning and national reconstruction."[29] Most editions of the journal were compilations of papers on a particular topic such as agriculture, engineering, or practical training that had been discussed at one of the national reconstruction forums. Papers ranged from think pieces in which the authors offered their perspectives (always well grounded in expert knowledge of the subject) on how China should best grapple with a reconstruction problem, to surveys of particular industries or products, to concrete descriptions of industrial or agricultural practices that the author had learned about while undergoing training. In addition to offering broad ideas about how Chinese development should proceed, therefore, the journal actually did contain information that might have been useful to people in China with a need for specific technical knowledge about subjects such as alcohol or penicillin production or dam construction. As Meng noted in the foreword to the practical training volume, "This issue contains brief statements by some of the Chinese student trainees. Through this Journal the writers make valuable contributions in sharing their experiences. Readers of this number in China will get some idea of what their fellow students are doing in this country, and be benefitted by their experiences; they may feel free to express whether further publications of such material will be useful as well as to make other suggestions."[30] Meng wanted both the forums and the journal to be part of a conversation with scientists and technicians in China and to be responsive to their needs.

Chih Meng and the China Institute thus supported Chinese students in the United States in a variety of ways and also helped mobilize those students to serve the Chinese nation. By seeking out and then distributing funding from both the U.S. and Nationalist governments, Meng was able to help students already in the United States survive the war years. The funding as well as other encouragement from Meng and the Committee on Wartime Planning reminded students of their patriotic duty and led many to pursue studies or training in fields that planners believed would be of concrete use to China in the short and long term. He further provided both in-person and print venues in which students could share their experiences and knowledge with others, develop networks, and become further indoctrinated in an ideology of national service. Not all Chinese students in the United States participated in these activities of course, but Meng and the China Institute made a point of tracking them all and the China Institute maintained the most comprehensive records of Chinese students in the United States throughout the war years and beyond.

Technical Training in the United States

Chinese students were not the only ones traveling to the United States to acquire knowledge to take back to China. During the war years, numerous practicing technicians participated in training programs set up by both governments as well as Chinese groups in the United States and some U.S. industries. On the Chinese end, these programs were for the most part devised and arranged by the NRC, the Ministry of Economic Affairs (MEA), the Ministry of Education, and the MOAF or by military organs such as the Commission on Aeronautical Affairs and the Directorate of Ordnance. These entities worked in concert with the U.S. State Department or other U.S. government organs such as the Department of Agriculture, Department of the Interior, and the Foreign Economic Administration (FEA), and a tight-knit set of Chinese organizations in the United States that included the China Institute, the Chinese Institute of Engineers, America Section (CIE), and the Universal Trading Corporation (UTC). Training programs, as distinct from study abroad, were short term and targeted at specific development goals that were typically determined by the Chinese government. They were not entirely new to the wartime period—the NRC had previously sent small numbers of technicians abroad in selected industries of high importance—but the scope of such programs dramatically increased during the war.[31] Participants sent from China were, for the most part, already employed in a government ministry, bureau, or enterprise and had been selected for participation through exams and because their areas of expertise were of critical importance to the implementation of Nationalist economic development plans. Some participants, however, did not fit this description.

CHINA INSTITUTE TRAINING PROGRAMS

Not all training programs were designed for technicians sent from China. Chinese students, recent graduates, and professionals already in the United States also undertook training for the purpose of better equipping themselves to serve the Chinese government. As with students engaged in academic study abroad, the China Institute and Committee on Wartime Planning played a major role in facilitating technical training for Chinese already in America by doling out scholarships and working with both U.S. government agencies and U.S. industries to place trainees. In addition to making a survey of technical personnel in the United States,[32] the China Institute facilitated postgraduate training in American industries for students already in the United States.

In his February 1945 semiannual report to the Committee on Wartime Planning, Meng noted that there were 654 Chinese in training programs in the United States. Many of them had originally come to the United States as students and others were Chinese Americans whom Meng had recruited into training programs that would prepare them to travel to China to participate in specific war-related projects. Of these trainees, 273 were in engineering, 110 of whom were in mechanical and aeronautical engineering; 130 were in the natural sciences including medicine and pharmacy; 83 were in accounting, economics, banking, and business administration; and 253 of the trainees were placed in defense industries including aircraft, automotive, rubber, chemicals, machine tools, and shipbuilding.[33] The China Institute arranged for industrial placements with General Motors, General Electric, Westinghouse, and Standard Oil, as well as in government offices including the Office of War Information, Board of Economic Warfare, Bureau of Agricultural Economics, Office of Education, and Bureau of the Census. In addition to providing technical training in specific fields, these placements gave trainees "opportunities for observation and participation in the wartime activities of the American government."[34]

As can be seen in the example of Chia Wei Chang, Meng was intimately involved in every aspect of the placement process. He received and dealt with petitions from students seeking placement and participated in negotiations with potential placements. Chang was a soil scientist who had received his BS from Jinling University in 1931, his MA from the

University of California in 1937, and was doing additional coursework and research under Emile Truog at the University of Wisconsin in 1942. By the spring of 1942, Chang had run into some unexpected difficulties in the United States. He had been ill, his family had unexpectedly arrived as war refugees from China, and he was, as a result, in some financial distress, now having to feed four mouths instead of just one. In search of a solution to his problems, he sought placement (and financial assistance) from Meng. Meng's response was to introduce Chang to Walter C. Lowdermilk of the U.S. Soil Conservation Service, who made arrangements for Chang to participate in fieldwork with the Physical Surveys division of the Soil Conservation Service. Lowdermilk, who himself was a great advocate of the importance of good surveys and maps to the development of a soil conservation plan, proposed that Chang take "practical training in the making of land use capability maps," a proposal that Chang enthusiastically accepted.[35] Although Lowdermilk had at the time already committed to travel to China under the auspices of the cultural relations program, Meng's connection to Lowdermilk likely derived from the fact that Lowdermilk had spent time in China in the 1920s and was well known in Chinese soil-conservation circles. Perhaps it was that experience that led Lowdermilk to actively partner with Meng, exploring opportunities for placement of Chinese trainees in various units of the Soil Conservation Service scattered around the country. Not all of Meng's American helpers had spent time in China. Some, such as L. A. Wheeler, director of the Office of Foreign Agricultural Relations of the U.S. Department of Agriculture, partnered with the China Institute because such activities were expected of them as part of their jobs.[36]

Meng also received suggestions for training programs and requests to round up trainees for them. One such suggestion came from Lowdermilk, who was, in the spring of 1942, preparing to travel to China as one of the State Department's cultural relations program experts. Lowdermilk's experience in developing a national soil-conservation program for the United States convinced him that aerial photography was the most expedient way to make the kinds of maps that would be required to develop a national conservation program. Such activities could not easily be undertaken during the war, but Lowdermilk, like many Chinese planners, did not feel constrained to provide only advice that would be useful during the war. To that end, he devised a plan to repurpose military airplanes and pilots after the war to undertake that photography and, in anticipation of that moment, enthusiastically proposed to train twenty Chinese engineering graduates in "the making of maps from aerial photographs so that on their return to China our program of land conservation, including soils, crops, waters and forests, may be facilitated even as it was in the beginning of our nation-wide program of land conservation."[37] Lowdermilk suggested to Meng that the twenty graduates be trained in the offices of the Soil Conservation Service in Washington, DC and their regional offices so as to accommodate the large number of trainees, noting that the service was already prepared to undertake the training.[38]

One targeted short-term training program that aimed to prepare Chinese already in the United States to serve immediate Chinese needs was the China Institute's Burma Road program. As Meng said of the program, "We were called on to help solve two serious problems: 1. The road suffered wash-outs and needed some kind of binder right away; there was neither time nor resources to pave it or to bring in building materials from outside. 2. The trucks imported from the United States were breaking down on an average of fifteen days' use; something had to be done quickly to stop the breakdowns and to provide enough repairmen." To remedy these problems, the China Institute actively recruited trainees to participate in two "crash programs," one in highway maintenance and the other in automotive maintenance. He also recruited civil engineering students to experiment with "Chinese binding materials to stabilize the surface of the road." Because the automotive maintenance program "called for far more men than the few mechanical or automotive engineering students available," Meng noted, "I traveled around the Chinese communities from New York to Detroit, enlisting over four hundred American-born Chinese of high school level and placing them in the factories of Ford and Chrysler to undergo intensive training for auto-mechanics. More than half of the graduates went to China to serve during the war."[39]

The Burma Road training program was unusual among overseas programs in that it aimed at producing middle- and lower-level skilled labor who would operate under someone else's direction. Meng wrote that as he designed that program his "eyes opened to the fact that industrialization needs a chain of trained personnel, from the skilled laborer, the foreman, the engineer, and the planner to the researcher. China had been training mainly the top echelons; hence, missing links were causing serious difficulties. For example, the breakdowns of trucks on Burma Road were due chiefly to lack of skills in loading and maintenance which were not taught in colleges and universities. This awakening led me to inaugurate an integrated program of technical training which included all branches of technology and which would train secondary as well as college students."[40] Although Chinese planners also discussed the need for a more integrated approach, they typically looked to the Ministry of Education to provide solutions to that problem by developing new curricula in vocational education or by obligating students in particular courses of study to undertake national service after graduation. Meng's approach was more similar to the Bailie approach as well as training programs in NRC factories, which also aimed at training unskilled labor.[41]

THE CHINESE INSTITUTE OF ENGINEERS
AND TRAINING

Like the China Institute, the Chinese Institute of Engineers, America Section also played an organizational role with respect to trainees. Although the CIE did not, for the most part, take on the responsibility of finding placements for trainees, it did reach out to newly arrived engineers and brought them into the organization so as to build networks and mutual understanding. The CIE had fallen into a state of inactivity prior to the war, but in 1942, a group of members led by L. F. Chen decided to resuscitate it by holding regular meetings and embarking on the regular publication of the *News Bulletin* and eventually a journal. The CIE's stated aims were improvement of engineering knowledge, cultivation of contacts, and cooperation with organizations in China. However, records of its initial meeting indicate that members also felt that the CIE should help arrange training and employment for members, offer advice on the field, and arrange for factory site visits in related industries.[42] Members included working professionals, engineering faculty, students, and short-term trainees.

Although the CIE clearly served as a community to foster networking among Chinese engineers in the United States, it also served as a conduit for news from home and through which its membership could feel that they were contributing to Chinese reconstruction. As one member wrote, "We engineers have pledged ourselves to concentrate our efforts on the development of post-war industries in China."[43] The CIE partnered with the UTC (a Chinese state-run company located in New York that will be discussed in detail below), the China Institute, and the Wartime Planning Committee, and both the main chapter and the local chapters of the CIE frequently held joint sessions with the national reconstruction forums at which the discussion topic was often related to planning for the industrialization of China. The *CIE Journal* published the proceedings of at least some of these meetings along with other papers submitted by the CIE membership with an eye to both making U.S. technical information available to China and introducing American readers to innovations in China.[44]

The bimonthly *CIE News Bulletin* printed updates on the activities of both the central chapter and the numerous local chapters that were established over the course of 1943 and 1944. It also contained news items about CIE members, editorials on various engineering-related topics, and news about engineering activities in China. In addition to these items, the bulletin published information about new engineering trainees arriving from China. At first, because the number of trainees was relatively small, they were identified by name and field and sometimes also by institutional affiliation in China. By the time the war was drawing to an end, however, the number of trainees had grown so much that lists simply included the number of trainees and fields of study. The bulletin kept track of only engineering trainees, who were mostly sent by either the NRC or a military entity such as the Commission on Aeronautical Affairs or the Directorate of Ordnance with the purpose of training them up in specific fields related to either industrial or military development. The CIE invited these trainees to join the America section for the duration of their stay in the United States, and branch chapters also frequently invited trainees stationed in their areas to present on conditions in China at branch meetings.

Although most of the CIE membership was, by definition, already at least partly trained, training was a subject of regular discussion in the bulletin. A common refrain was concern that students and trainees get the right kind of training. As L. F. Chen observed at the inaugural meeting of the newly reconstituted CIE, trainees should take care to avoid ending up as "Jacks of All Trades" who would be "incapable of offering the kind of specialized service that China sorely needs."[45] Both Chen and later contributors to the bulletin frequently observed that it was not uncommon for engineers in China to be required to serve as administrators as well as technical specialists; for this reason, the CIE regularly urged trainees and students to study administrative sciences.

The bulletin also served as a place for U.S. corporations to advertise practical training opportunities, often with an eye to killing numerous birds with one stone. In the first place, under wartime conditions during which American recent graduates and students were subject to the draft, companies were in need of educated workers whom they could train relatively easily, and Chinese engineers fit the bill. Second, many of the companies that advertised training programs were interested in doing business in China after the war. Chinese wartime trainees could assist with this endeavor, either as employees or as friends of the company once they returned. Third, by establishing such training programs, these companies contributed to the Allied war effort. Finally, as an article titled "Chinese on the Job" observed, U.S. companies often found Chinese workers to be model employees and thus especially liked to hire them.[46] It served the economic interests of U.S. companies to offer training opportunities to educated Chinese engineers. With the assistance of skilled Chinese workers, they could not only maintain operations and continue to produce in a time of labor shortages, but they could also build relationships that might later prove to be helpful in their efforts to expand their markets after the war.

Another training issue taken up in the pages of the bulletin was the need to organize a training plan for students and engineers already in the United States that would provide them a pathway to contributing to postwar reconstruction. Concerned that the Chinese government was only planning for the utilization of trainees that it was selecting and sending from China, the editor of that issue of the bulletin (Wei-pang Han, a leather expert working with the UTC) argued that "there are still hundreds of Chinese engineers and engineering students in the U.S. who have as yet no connection whatsoever with the Chinese Government organizations or private industries. Many of them are at present engaged in essential war work, but while doing their bit for the Allied cause, they are frequently engulfed in uncertainties regarding their future course. Anxious to do their best in one way or another, to help in the forthcoming industrialization program, yet without the sense of being a member of a team, they tend to build up their technical experience in a haphazard pattern." A real training program should be designed, he proposed, to guide these engineers in their development so as to prepare them to be of the greatest possible use in China after the war. Once designed, the program could be facilitated by the CIE and the China Institute who could help to identify suitable placements for trainees.[47] It is unclear whether Han was speaking for himself or if he was acting on behalf of his government to try to rally Chinese engineers already in the United States to connect with the Nationalist government and call for the development of such a program. Either way, no central plan was ever devised that incorporated Chinese already in America, though as we have already seen, the China Institute did take it upon itself to orchestrate training opportunities for Chinese in America that it, in conversation with Chinese leaders, deemed to be of potential utility.

THE UNIVERSAL TRADING CORPORATION

Yet another institution that facilitated training and the transfer of technical knowledge from the United States to China was the UTC (世界貿易商業), which was set up by the Ministry of Finance in 1939 for the purpose of managing sales of tung oil and other Chinese products and making purchases for the Chinese government.[48] The UTC was in charge of managing a $25 million loan that the U.S. government had granted with the expectation that China would repay it with profits from the sale of tung oil in the United States during the war. The loan was granted for the purpose of enabling China to purchase American manufactures that could be used for national defense but not actual weapons or other military equipment.[49] The UTC's most basic function, therefore, was procurement. It engaged directly with U.S. companies to purchase U.S. manufactures for China, and as such, it played a major role in opening direct pathways between those industries and their client, the Nationalist government. The U.S. industries with which the UTC traded thus had a clear economic interest in fostering a relationship with China. The UTC's staff of engineers, many of them U.S. trained, also assisted with "obtaining designs for various projects including chemical, mechanical and power plants and in securing information to help solve specific production problems."[50] To do this, the UTC sometimes negotiated technology transfer contracts with U.S. companies such as Westinghouse.[51] The UTC was thus tasked with utilizing the relationships it established through trade as pathways to acquire knowledge. Although they sometimes had to pay for that knowledge in consulting fees, they also took advantage of those companies' need for labor in conjunction with the China Institute training programs to place trainees in key industries.

Working closely with both the Chinese Institute of Engineers and the China Institute, the UTC set up industrial training placements in the United States for technocrats working in the enterprises of the NRC, NBIR, and others. It also fostered the work of the China Institute, the Wartime Planning Commission, and the Chinese Institute of Engineers by facilitating their communications with the Chinese government and with researchers in China and by providing New York office space to the CIE. The UTC maintained its own staff of engineers who were "on the spot and in close touch with the general industrial conditions in the United States and who" by 1943 had "already familiarized themselves with many standard and emergency products as well as facilities" and were "in a favorable position to observe and analyze the developments that are taking place here [in the United States] and to make proper selection of technical facts and methods as well as general information to be reported to the industrial executives and engineers in China."[52] Among the engineers attached to the UTC in 1943 were S. Y. Ma (plastics and fertilizers—a colonel and a chemical engineer), K. J. Kang (automotive engineering, roads and fuel), William Yep (gasoline refining), T. H. Ku (power), L. P. Yeh (radio and infrared), T. P. Hou and W. Y. Pan (coke ovens), C. L. Shen (gauges), U. K. Chan (power), C. Y. Wang (metals), J. H. Tsui (electricity), C. H. T'ang (dams and water management), W. P. Han (leather), C. S. Chen, (factory organization), C. S. Kwan (factory organization), H. M. Yu (automotive), C. L. Tsung (electricity), Wu Liao (television), and H. C. Cheng (chemicals). Some of these men were involved in the CIE as well.

At least some of the UTC engineers were fairly high-profile figures who, in addition to engineering work, also participated in national reconstruction forums and other such activities. One of these men was T. P. Hou 侯德榜, a chemical engineer and expert on the manufacture of soda who had, prior to the war, been the vice-president and chief engineer of the Yongli Soda Plant in Tianjin. Founded in 1917, Yongli was a private enterprise. After the war started, the management of Yongli refused to collaborate with the Japanese and refocused its efforts on developing the chemical industry in inland China. As it did so, it found itself in a contentious relationship with the Nationalist government, receiving financial support for the critical industries it was developing but simultaneously pressured by the state to give itself over to state ownership. Not until 1952, however, did it become a state-owned enterprise under an entirely different government.[53]

As a UTC engineer, Hou was now, effectively, working for the government, but he continued to spend a good deal of his time working with his Yongli colleagues and other chemical engineers. Although he spent a good part of the war in New York, Hou worked from a distance with colleagues in Sichuan to develop a new process for manufacturing soda and ammonium chloride (both fertilizers).[54] In addition to these scientific activities, however, Hou also worked to facilitate KMT goals in New York (though unsurprisingly, his views on nationalization of industry were not entirely in sync with those of KMT leaders[55]). He served on the board of the Committee on Wartime Planning as well as on the CIE's Special Committee for Professional Consultation, set up in 1944 to give professional advice to CIE members.[56] In his role on the Committee on Wartime Planning, Hou gave presentations at various national reconstruction forums and also edited the third volume, on engineering, of the *National Reconstruction Journal*. P. W. Tsou 鄒秉文, an agriculture specialist who was also closely associated with Yongli Chemical Industries and spent a good deal of the war in the United States advocating for agricultural cooperation between the United States and China, also had at least some relationship with the UTC as he participated in a UTC-sponsored trip along with several others to tour the TVA and its related enterprises in December 1943.[57]

The NRC also set up its own technical group in May 1943 that would be located in the United States and perform some of the same functions as the UTC engineers. The NRC Technical Group's work, however, was more explicitly aimed at responding to specific assignments from the NRC that connected to postwar industrial reconstruction. The group's organizational charter indicated that its role was to investigate, seek advice on, and conduct negotiations on "any assigned subject" and that the members of the group would be selected in "regard to the mapping out of certain post-war industrial reconstruction plans."[58] This group also conducted site visits to factories, consulted with American engineers, and generally collected information to send back to China. But whereas the UTC was primarily concerned with helping to solve immediate problems, the NRC Technical Group appears to have been more explicitly focused on postwar planning.

The UTC also worked with Chinese government organizations to set up training programs for their staff. In 1942, for example, S. D. Ren, vice-president of the UTC, arranged with Y. T. Ku, head of NBIR, to bring a group of twelve NBIR engineers to the United States to undertake practical training in their fields.[59] It seems likely that the UTC also helped to arrange for the placement of the numerous other NRC engineers who traveled to the United States before 1945 for practical training. However, by early 1945, if not earlier, the training group of the NRC in New York had entirely taken over this work.

With time, the UTC also took on an important educational role for Chinese engineers in the United States as well as those who remained in China. To this end, the UTC housed a technical library at its New York headquarters and published the *Universal Engineering Digest*, which, according to the *CIE News Bulletin*, served "to introduce to our executives and engineers in China, information useful in the solution of our country's war production problem and helpful in the preparation work for postwar reconstruction."[60] The digest aimed, in particular, to inform Chinese engineers of new ideas and methods that were being developed in the United States to meet war demands, stating in the foreword to the first issue that "these are things that will most surely stimulate the thinking and accelerate the efforts of the Chinese engineers and industrial executives who are tackling the same problems in China."[61] Starting in April 1943, the digest was published five times each year on lightweight paper suitable for airmail distribution to China (see fig. 3.2). The UTC also distributed the digest to Chinese engineers in the United States.

Written entirely in English, the digest contained condensed versions of articles originally published in a wide array of American engineering journals as well as new articles written by engineers attached to the UTC who were spending their time working in or observing American factories and familiarizing themselves with both the machinery and the organization of those factories. Some critical topics, such as the manufacture

FIG. 3.2. The front cover of the August 1943 edition of the *Universal Engineering Digest*. The image shows two Chinese trainees operating a machine at the American Machine Tool Company.

of aviation fuel, were covered in most issues of the publication. Other topics were aimed at specific problems that engineers faced in China. For example, the third issue of the journal contained articles on how to reduce waste of welding rods, recondition engine cylinders, and safely overload generators, all problems that engineers in resource-poor China might encounter.[62] The digest would also periodically focus an entire issue on a specific topic such as the auto industry or telecommunications. These special issues were written primarily by UTC engineers and were based on their observations of U.S. industry. A major aim of the publication was clearly to produce, for the consumption of Chinese engineers in China, technical information that could be concretely useful to them as they sought to solve specific problems in the extraction of resources and the manufacture of goods. Therefore, rather than focusing on a problem and offering a plan, which was the approach taken by most of the articles in the *National Reconstruction Journal*, digest articles typically described a machine and how it operated, or laid out the steps of a specific industrial process.

NATIONAL RESOURCES COMMISSION
TRAINING PROGRAMS ABROAD

Among the NRC trainees who came through the UTC were the so-called thirty-one, the first large group of NRC engineers to be sent to the United States (though the NRC's Central Electrical Manufacturing Works had sent a group of four trainees to the United States the previous year, in the fall of 1941). The thirty-one were selected by government examination as well as NRC evaluation of the contribution that they could make to preferred projects. To be eligible to participate in the NRC training, the trainees had to meet three basic requirements: (1) be graduates in industrial or related disciplines, (2) have good English and Chinese, and (3) have at least five years of work experience.[63] By and large, the trainees already had considerable experience working in their respective fields, though they were also relatively young. The thirty-one, who, with the exception of petroleum geologist Sun Jianchu 孫健初, were all between the ages of twenty-seven and thirty-seven, arrived in New York in a series of small groups. Among them were four petroleum engineers, six energy specialists, six electrical engineers, four mechanical engineers, four chemical engineers, three metallurgists, two coal specialists, and two specialists in management of factories and mines.[64] Aside from three of the petroleum engineers who were sent in May 1942, the other trainees all arrived between September 1942 and March 1943.[65]

Training was arranged for the group by L. F. Chen who, working with the UTC, organized an individualized series of field-appropriate on-site factory trainings for each trainee. For example, An Chaojun 安朝俊, one of the metallurgists, was sent to the U.S. Steel Corporation and the Carnegie Steel Corporation, both in Pittsburgh, for six months each, followed by another six months of site visits to various other foundries in Philadelphia, Cleveland, Detroit, Chicago, and New York.[66] With the exception of a couple of the petroleum engineers who were engaged in academic study, all of the trainees had similar arrangements, most undertaking initial training in a single factory for a duration of four to twelve months and then moving on to other factories for shorter stays. The engineers were supported by remittances from the NRC, though many also earned salaries in the plants in which they were working (archival documents suggest, however, that the total in earnings and remittances was still not sufficient to cover their expenses).[67]

Throughout the trainees' time in the United States, the UTC continued to facilitate placements and UTC engineers acted as intermediaries between the trainees and the companies with which they would be placed. L. F. Chen of the NRC in New York, however, was the primary point of contact for the group and he corresponded regularly with all of the thirty-one engineers about their placements and their experiences. For example, he kept in close contact with Y. S. Sun 孫運璿, a hydroelectric power specialist who spent his initial training time at the TVA, receiving frequent reports from Sun and writing prompt and appreciative thank-you notes to the various plants that Sun visited. Chen negotiated with the Pacific Power and Light Company on Sun's behalf, arranging for his training there following his time at the TVA and also received a very favorable report on Sun's training from George Bragg, the vice-president and general manager of that company.[68] Some placements were facilitated by Americans who had spent time in China working on NRC projects either as part of the cultural relations program or in other contexts. For example, Chen worked through contacts given him by Paul B. Eaton of the State Department's cultural relations program to arrange for trainees to work in Bell Telephone

Laboratories.[69] The trainees themselves also made their own arrangements to meet up with American contacts. Sun Jianchu of the Yumen Oil Field, in the United States to survey oil production facilities, met up with M. H. Bush while visiting the General Petroleum Corporation in Bakersfield in December 1943. Bush had spent some time as a consultant at Yumen, helping with the development of the oil field earlier that same year.[70]

In the spring of 1942, before most of the engineers departed for the United States, the NRC held a series of meetings for the thirty-one in Chongqing. At least one aim of the meetings was to help the group develop a sense of camaraderie and familiarity so that once they got to the United States, they might continue to work together for common ends. Although they were scattered across the country at an array of different industrial sites, the thirty-one did get together for meetings while in the United States. Every six months they had a meeting orchestrated by NRC representatives in the United States; at other times, some members of the group held their own technical meetings at which they discussed their work and through which they maintained close contact and coordinated their efforts. Their first biannual meeting was held in New York on January 1–3, 1943. It was presided over by L. F. Chen and Yin Zhongrong 尹仲容 (also known as K. Y. Yin) of the UTC, at the New York NRC office.[71] The second biannual meeting was held in Knoxville, Tennessee on July 21–25, 1943, perhaps signaling the importance of the TVA training to the NRC, which was looking to develop its own Yangzi Valley Administration in China. Starting on April 4, 1943, regular technical meetings were also held near the TVA and were attended by whoever could get to them (not only the engineers studying at TVA sites).[72] These meetings tended to attract from ten to twelve participants and were held every two months or so. They involved a combination of technical reports by the NRC engineers and site visits to TVA industries near the meeting locations. The trainees wrote up reports on their technical meetings, which they submitted to both the NRC and the China Institute.[73] At the third technical meeting, held August 13–15, 1943, in Chattanooga, the twelve participants resolved to start preparing papers ahead of future meetings that could form the basis for technical discussion and exchange of ideas, and to start a series of chain letters through which technical discussion could be carried on even when the engineers were not together.[74]

As of May 1944, only three of the thirty-one had returned to China, four had arranged to extend their trips for six months, and the remainder were on schedule to return at some point in 1944 in accordance with plans that had been made at the time of their departure. Despite a request for thirteen of them to remain in the United States to continue training so as to better prepare themselves to assist in China's postwar reconstruction, Weng Wenhao, minister of economic affairs and head of the NRC, held firm, insisting that they return at the end of their scheduled training periods and informing their work units that they should start planning for how best to make use of the returnees.[75] L. F. Chen notified trainees of this decision, alerting them that in order to extend their stay, they would "have to give full details and reasons" and suggesting that training extensions might be difficult to get.[76]

Both during the course of their training period and as they were returning to China, each of the thirty-one wrote up extensive reports describing their training program and what they had learned. Some of these reports were in English, others in Chinese, some filled with technical detail, others of a more general nature. By and large, however, the reports went into considerable detail, describing organizational systems in factories and larger entities such as the TVA, providing data on the production and operations of plants, dams, and other things, drawing diagrams of machinery, and providing chemical diagrams, relevant equations, and descriptions to explain how the machinery worked, expected chemical reactions, and so on.[77]

As the war progressed, the NRC continued to send new groups of trainees to the United States. For example, in the late spring and summer of 1944, a new group of thirty-three trainees from various NRC enterprises was sent for training in the United States.[78] As 1944 was drawing to a close, L. F. Chen, who was increasingly devoting his time to arranging technical training, was anticipating the arrival of a group of ninety-nine engineers in February 1945, with another large group to follow in the summer of 1945.[79] By the end of the war, the thirty-one would look like a fairly small group as the number of trainees sent by the NRC increased dramatically.

Over the course of the war, Republic of China (ROC) leaders in the Ministry of Education and the NRC increasingly advocated for the training of large numbers of technicians in the United States, much of which would be financed, they hoped, by lend-lease funding. By the latter part of the war, bureaucrats in both institutions had their sights set squarely on the requirements of postwar economic development, and the availability of lend-lease funds made it possible to carry this out. Already by the summer of 1943, the Ministry of Education was reported to be developing a plan to send as many as one thousand students a year to the United States and Great Britain (seven hundred to the United States and three hundred to Great Britain) of whom half would be selected by examinations and half would be appointed by institutions. According to George Atcheson, the U.S. chargé in Chongqing, "The project has arisen from the lack of trained technical personnel in China and the anticipated needs for the post-war reconstruction program." He went on to note that this plan represented a reversal of earlier ROC government policy in the early war years, which forbade students and technicians to study abroad because their services were needed in China, but in his view, the wartime deterioration of Chinese universities meant that "the caliber of Chinese graduates was lowered and trained technical personnel became inadequate."[80] Weng Wenhao expressed essentially the same concern a couple of years later: "that Chinese education had suffered seriously from the war and that the teachers had been treated very poorly . . . [and] that even at its best Chinese education had not been adequate from the technical point of view."[81] But the problem was not just with the quality of education in China. Industrial work environments were, for the most part, so makeshift and so far removed from modern industrial shops in places like the United States that although engineers may have had frequent call to use their training, creativity, and ingenuity to bring inadequate machines up to par, they had no opportunity to learn how to operate in factories that met the highest industry standards. Inadequacy of local educational and training opportunities, combined with a desire to improve Chinese industrial productivity, meant that the Chinese government needed to find alternative strategies for educating technicians, and sending large numbers of trainees abroad was one way of achieving this end. Over the course of 1943, the plan was refined in various ways. By December 1944, the proposed training program had morphed into a plan to train 1,200 Chinese technicians (rather than students) with lend-lease funding.[82] But in spite of Weng's compelling arguments, even as late as December 1944, U.S.

government representatives expressed concern over the departure of so many skilled technicians from China "now when their services are urgently required in China"[83] and urged that the project be put on hold or scaled back. From the U.S. perspective, China should be employing all of its technical experts to help with the war effort, not training them for postwar reconstruction.

In spite of these concerns, the project did move forward. By early 1945, the CIE was gearing up to assist in the welcoming and placement of 1,200 trainees of whom they expected 1,011 to be engineers. In anticipation of their arrival, the CIE reorganized its membership committee so as to more efficiently connect with the new arrivals and process their membership applications.[84] The training was to be paid for by lend-lease funds and jointly administered by the training department of the NRC-NY, and the China Supply Commission (attached to the Chinese embassy in Washington, DC) and the International Training Administration in Washington, DC. As of May 22, 1945, Chinese government representatives in the United States had signed a contract with the International Training Administration to assist with administration of the training and orchestration of placements at a cost of $18.25 per trainee per month. Seven hundred and twenty of the trainees arrived before the end of the war, though the greatest number of those (629) did not reach the United States until the summer of 1945.[85] Details of this program are discussed in chapter 5.

MILITARY TRAINING

The NRC was not the only Nationalist government institution to arrange training in the United States during the war. Other government units, operating either in response to immediate wartime need or in an effort to implement long-term planning, developed their own training programs. For example, well before the NRC started sending trainees in 1942, the Chinese military had set up overseas training programs using students who were already abroad and already connected to the military. One such program was in aeronautical engineering.[86]

Already in the 1930s, recognizing the importance of developing its own aviation industry and military aviation program, the Nanjing government oversaw the establishment of a set of new institutions, governing structures, and also a new academic program in aeronautical engineering. Zhongshan University's eighteen-month program, which was quite innovative among Chinese academic programs of the time, consisted of four terms' worth of coursework and a two-month factory internship at the Jianqiao Aircraft Factory in Hangzhou.[87] Following the graduation of the first group of Chinese-trained aeronautical engineers from their program at Zhongshan University, several students took examinations for scholarships that would permit them to study in the United States at MIT. Among these early aeronautical engineering students was Yaozi Li 李耀滋, who had been one of twenty-one students in Zhongshan's aeronautical engineering program in 1935. Li received a scholarship to MIT just as the war against Japan broke out; by 1938, he had completed his master's degree at MIT.

The urgency of war, combined with requests coming from the Nationalist government or colleagues back home, however, compelled at least some students already in the United States to abandon more theoretical studies in favor of more active projects that had direct application to the war. Although some of Li's classmates went on to become professors of aeronautical engineering in both the United States and China, Li himself was persuaded that he should abandon his studies and apply his knowledge for the benefit of his country. In the summer of 1938, he was asked by China's Aviation Commission to travel to the Curtiss Airplane Factory in Buffalo, New York to inspect forty P-38 airplanes that China had ordered. Since the Jianqiao Factory at which Li had done his internship in Hangzhou had been a joint venture with Curtiss, Li was particularly well positioned to perform this work and also to get the most out of this experience.[88] By 1939, H. C. Qian 錢學榘, one of Li's former classmates in the Zhongshan program, had asked Li to give up his doctoral studies in favor of returning to Buffalo to take possession of and disassemble a set of machinery from a truck factory that the NRC had purchased and planned to ship to China. According to Li, Qian, who was himself already in Buffalo managing the disassembly process, argued that "resistance against Japan was every Chinese person's responsibility. Studying for a Ph.D. degree," he said, "would usually take several years, and was not the most urgent priority during the war. He asked me to give up my Ph.D. candidacy, and go to Buffalo to help him, as soon as I could."[89] Li went to Buffalo along with a number of other Chinese students from MIT. After that project was completed, he went on to work for B. L. Li of the Chinese Aviation Commission to help him negotiate a new joint venture with the Wright Aeronautics Company through which a group of Chinese aeronautical engineers would first train with Wright in New Jersey and then return to China along with Wright engine parts to assemble engines in China. Among those trainees were some of Li's classmates from the United States along with other engineers sent from China by the Aeronautical Commission. Li and the others spent the first part of 1940 training at the Wright factory and returned to China in October of that year to establish a new aircraft engine factory.[90]

The new assembly factory in Crow Cave at Dading in Guizhou was established in January 1941 and became at least partly operational by December 1942. Li, Qian, and others, many of whom had returned from study in the United States, led "a bunch of newly recruited young engineers to open up the wilderness and create a modern plant."[91] The plant was unusual among Chinese factories in inland China in that its machinery was almost entirely U.S. made and the plant operators were using U.S. specifications to manufacture airplane engines. Even so, a good deal of ingenuity had to be put into getting all of the imported machinery to work in a power- and raw material–scarce environment. The young engineering graduates at Dading underwent on-the-job training as they got the plant up and running, but it was not until Chiang Kaishek visited the factory in March 1943 that, at his suggestion, a formal training program for aircraft manufacturing technicians was established. Beginning in 1944, the plant started recruiting high school graduates to train in airplane engine manufacturing. By the time the plant packed up to remove to Taiwan in 1949, eight groups of trainees had passed through the program. One trainee was Wang Wenhuan 王文煥, who continued to work at the plant after his training and then moved to Taiwan in October 1949 along with about one-third of the plant's personnel and 5 percent of its machinery. Even though the plant was not really productive until after the end of the war (and then was at least partly converted into an auto parts factory) and thus failed to achieve its mission of "saving the nation by air," Wang later observed that it played an incredibly important foundational role in the development of the aircraft industries in both China and Taiwan because of the large number of trainees who passed through the program and worked there.[92]

In 1944, the Commission on Aeronautical Affairs undertook to develop a more extensive training program in the United States. One participant in this program, who went as part of a group of ten trainees from the Dading plant, was Wu Daguan 吴大观. Wu had graduated in aeronautical engineering from Southwest University in Kunming in 1942 along with twenty-four or twenty-five other students, some of whom, like Wu, had then worked at Dading. Wu remained at Dading until his departure for the United States in October 1944. He and the other trainees, who were located at the Lycoming factory in Williamsport, Pennsylvania, were assigned to study different engine parts. According to Wu, some studied cylinders, others pistons, and still others crankshafts and rods. Although originally tasked to study design, Wu ended up spending a year studying gears because no one else had been assigned to that area. In total, the group, which eventually expanded to include twenty-five trainees, spent a year and a half in Williamsport studying both theory and practice of their particular engine parts before returning to China.[93]

The Chinese military had immediate needs and did not hesitate to recruit students in the United States to give up their studies and put their knowledge to work. But at least in the case of aerospace engineering, even though they were able to press students into service fairly quickly, it still took a good deal of time to get projects off the ground. The number of people in China with the necessary knowledge and experience was simply too small. Some knowledge could be built by technicians doing on-the-job training under the supervision of leaders at Dading who had already been trained in universities and factories in China and the United States. However, training abroad took time, as did the acquisition, shipping, and installation of machinery in Dading. The projects that the thirty-one were being trained for were similarly pressing, but their training programs were, in many cases, even longer. Even training programs designed to address immediate needs could not be completed quickly.

AGRICULTURAL TRAINING

As we saw in chapter 2, agriculture, the development and modernization of which required strategies for long-term and lasting change, was a matter of particular concern to some Chinese planners, including participants in the national reconstruction forums. One area that planners understood would be critical to the success of agricultural plans was agricultural extension. Without trained extension workers, the possibility of transforming the agricultural practices of China's farmers seemed remote. As P. W. Tsou wrote in 1944, China had "agricultural problems for which experts have already mapped out solutions, but the effect of such solutions has not yet been very extensive because of lack of the extension organizations to carry them out." Among the solutions that Tsou hoped could be implemented were improvements to rice and wheat output by planting new varieties of both crops that had been found to be well adapted to Chinese soils and climatic conditions, application of new fertilizers in accordance with the findings of scientific studies, new methods of controlling both insects and plant diseases that had been studied by entomologists and plant pathologists, and new disease prevention strategies for livestock that had been developed by scientists working for the MOAF.[94] According to Tsou, none of these new strategies and techniques were being implemented to the degree that they should be because of a lack of extension workers to disseminate the knowledge and encourage farmers to change their practices.

As part of his broader plan for agricultural improvement and development, Tsou recommended that the Central Agricultural Promotion Committee be reorganized as the Central Agricultural Extension Office and that provincial offices of agriculture and forestry establish provincial farms to grow seeds and undertake other extension activities. He further proposed that each county (*xian*) set up an agricultural extension office with at least one extension agent. Finally, he suggested that farmers organize themselves into village-level agricultural improvement societies. With this new structure in place, Tsou argued, extension work could succeed. "Extension agents of the county will be responsible for directing the extension work of their respective agricultural improvement societies. District extension offices will direct and help their respective county extension agents. Provincial offices of agriculture and forestry will direct and help district extension offices. Finally, the Central Agricultural Extension Office will direct and help provincial offices of agriculture and forestry."[95] Tsou further proposed that China's agricultural colleges train agricultural specialists to do extension work and that "tens of thousands" of extension workers be recruited to serve the various offices of the MOAF as well as the many county extension offices.[96]

Yang Shu-chia, who worked for a time for the China Institute, argued that for extension work in China to truly work, it would require the creation of agricultural cooperatives that would help with the dissemination of research-based knowledge and with convincing farmers to implement techniques based on that knowledge. "In China," Yang wrote, "if an extension service man speaks to 400 farmers a day his influence may not reach more than 1,600 acres [owing to small farm size]. Further, many of the farm audiences in China are literally half deaf either because they do not know what the extension man talks about or because they have no confidence in the new methods introduced by the speakers."[97] Training more extension workers, therefore, might help to a certain extent, but from Yang's perspective, additional structures needed to be put in place to make the most of extension work. Like Tsou and Yang, Yui Shin-min, a tung oil expert who was concerned with the development of a specific crop, also advocated for a comprehensive reorganization of China's extension service.[98]

National reconstruction forum participants were not alone in their concern over this issue. To address the deficit of highly trained extension workers, the Chinese government, working with the U.S. Department of Agriculture, laid plans in the summer of 1945 to send a group of extension leaders to Iowa State College for training in the fall of that year. Twenty-one students, at least one of whom was female, arrived in Ames in September for a nine-month training program. They remained in Ames for a short time to acclimate themselves and were then sent off to other sites for three months. Each student was to spend one month on a farm, another with county extension agents, and a third with state extension specialists. In January, they returned to Iowa State College where they took specially developed training courses until March, after which they were sent out again for three more months of practical training in other states and one month with the Department of Agriculture in Washington, DC.[99]

Instructors for the program were faculty and staff of the Iowa State College extension service including E. F. Graff, district extension supervisor, William Homer Stacy, Iowa State College extension sociologist,

J. W. Merrill, district extension supervisor, and Professor Murl McDonald of the extension department. Instruction focused on selection and training of personnel, leadership training and selection, and relationships with agencies and organizations. Graff, writing to Stacy about how the early part of the course was going, indicated that he had "endeavored to get as much discussion out of the group as possible based on their field observations."[100] Stacy taught the course on leadership training and selection and aimed to teach trainees about the importance of voluntary leadership, focusing on how voluntary leaders should be recruited, trained, and directed. For extension work to succeed, according to Stacy, it needed "to reach more people, to make the program the people's program, to help the people develop themselves," all of which were goals that made sense in the Chinese context. To accomplish these goals, Stacy believed that extension agents needed to identify and work with "local men of prestige, paid workers in related fields, [and] unpaid volunteers."[101] Trainees were taught how to interact with local government and how to develop a calendar that would keep volunteers engaged. On the assumption that data collection was an important dimension of extension work, they were also instructed in survey methodology, including how to train survey workers in interview methods and working with forms. The curriculum was very interactive and made use of teaching materials that had clearly been developed for use in similar training programs for American extension workers. Stacy taught the trainees that "postwar planning in the average community means: 1. Having a few leaders 'talk it up' with others. 2. Gaining the support of the town council and other responsible groups. 3. Holding a community planning conference bringing together leaders of all institutions and organizations. 4. Establishing a Community Coordinating Council to correlate and carry forward the program. 5. Designating community action committees to obtain facts and to develop projects and programs."[102] Although not all of these ideas matched up perfectly to conditions Chinese extension workers would have found back home, they would nonetheless have provided a foundation for students grappling with the problems mentioned by Tsou, Yang, and Yui.

A year after the trainees had returned to China, Graff sent copies of a term paper written by one of the trainees to the other instructors of the course as an example of precisely the kind of thoughtful engagement with the subject matter that he had hoped to see from students. Based on the paper, and presumably also on his overall sense of the course, Graff wrote that "I think the Iowa Extension Service made a fine contribution to International good will in taking adequate care to give good training to these Chinese people."[103] In his term paper on "Principles and General Techniques of Effective Teaching," trainee Sam H. K. Shih wrote that the extension students put "painstaking emphasis on teaching method" because "we are concerned mainly with teaching method and educational process. If we want to 'Extend' we must sale ideas first. For instance, Miss A, specialist authority in food nutrition, when she gave a radio talk, she talked on technical terms. She did not know how to make her audience to understand her teaching. She fail in arouse interest, she cannot convince others. Her ideals can't sale out, she can't 'Extend.' We Extension worker, though not know very much about food nutrition as Miss A does, but she has to ask our help (or follow our teaching principles) in order to extend her ideas & change farmers behavior."[104] Although Shih did not express his ideas in perfect English, it is clear that he well understood the importance of effective communication in extension work. As he observed, "We can't help the farmers to develop themselves without arousing their interest."[105] Farmers must have interest in order to want to learn, and in order to make extension projects work, extension workers needed to be able to stimulate continuing interest. In his previous extension work, Shih said, "I could not gathered 1/10 of the farmers to come to the ext. meeting in my county. I almost disappointed that farmers have no needs & interest."[106] Finally, he discovered that if he announced that he would be showing a film, everyone from miles around would show up, but their presence still did not guarantee sustained interest in his projects. At Iowa State, Shih felt, he had learned educational strategies that he would be able to use to generate that kind of interest in the future. As a trainee with prior experience in the field, Shih was particularly well positioned to get something out of the training he received. He already understood the challenges of extension work and seemed eager to overcome those challenges. His training program provided strategies that might help him do that.

Extension training was but one of many agricultural training programs for Chinese in the United States during and immediately following the war. P. W. Tsou was not just a planner; he was a doer, and his mark can be seen on virtually all proposals for training Chinese agricultural specialists in the United States in the latter part of the war. In the summer of 1944, as part of his broad plan for improving Chinese agricultural productivity, Tsou presented a request to the International Harvester Company, a company that had been selling agricultural implements in China since 1918 and that had already demonstrated its commitment to the improvement of Chinese agriculture through its support of James C. Yen, pioneer in mass education and rural reconstruction. By January 1945, International Harvester had announced a new program to train twenty Chinese students in agricultural engineering at Iowa State College and the University of Minnesota, crediting Tsou with the idea.[107] According to G. C. Hoyt, vice-president of International Harvester, "Sound agricultural development in any country requires the leadership and guidance of men who are trained in the specialized knowledge that is needed. It is the purpose of this cooperative program to provide that training. When these young men return to China they will be prepared to carry on research and development work in field equipment, farm power, rural industry, farm structures, applications of rural electricity, and soil and water management; or to serve as agricultural engineering instructors, or as extension service workers, taking information on improved farming practices to Chinese farmers."[108]

Hoyt had high hopes not only for the training program he and his company was sponsoring but also for the long-term results it would yield. In his vision, the International Harvester trainees would become leaders in their field who would bring modern, American methods to bear on Chinese problems. Unstated but certainly implied was that some of those methods would surely involve International Harvester products. International Harvester may have laid out a bit more cash on its training program than other kinds of companies that took in Chinese trainees during the war, but if through their program they could train more technicians who would become leaders (and influencers) of all dimensions of China's agricultural development, it would be worth it.

The twenty Harvester fellows were selected by evaluation of their academic records and examinations held in Chongqing, Chengdu, Kunming, Guizhou, Xian, Lanzhou, and Jianyang that were jointly administered by the Ministry of Education, the MOAF, and the Agricultural Association of China.[109] All had graduated from Chinese universities, ten in agriculture and ten in engineering. The first group of ten students arrived in the summer of 1945 and the second the following winter for a three-year training program. Agriculture graduates were sent to Iowa State, and engineers to Minnesota. Since the aim of the program was to produce a group of agricultural engineers and since none of the students had trained in that field, initial instruction focused on filling gaps in the undergraduate agricultural engineering curriculum. Students at Iowa State filled the engineering gaps, and those at Minnesota filled the agriculture gaps. Once the students had filled the requirements of an agricultural engineering undergraduate degree, they went on to complete MS degrees in agricultural engineering. As part of their training, the students were exposed to mechanized farm equipment on farms operated by their respective institutions.[110] "During summer vacations and between quarters, the students were given field assignments in a wide variety of subjects, including rice, cotton, tobacco, fruit growing, truck farming, irrigation, and farm management. Many had opportunity to work with Harvester dealers and sales branches as well as at Harvester and other factories producing agricultural implements." By early January 1947, the students, seventeen of whom had already completed their master's degrees, were embarking on a ten-week practical training course on an eighty-acre farm near Stockton, California. The course consisted of six hours of fieldwork and two hours of classroom instruction each day on "setting up, operation, adjustment, and maintenance of farm equipment." Following the training, the students were "assigned to farms, field experimental stations, and factories for additional work in their fields of special interest" until June 1948, at which time they would return to China. As of the winter of 1947, the plan was for the MOAF to assign the returned students "to carry on research and development work in field equipment, farm power, rural industry, farm structures, applications of rural electricity, and soil and water management."[111] International Harvester, of course, also hoped that as the trainees conducted that work, they would help to build a market for the company's products in China.

Conclusion

Overseas training programs designed to address specific development needs became increasingly popular as the war progressed and continued after the war ended. Such programs were typically the result of engagement between the Chinese government, its intermediaries in the United States, and U.S. universities, corporations, and government agencies. To help arrange these training opportunities, the Nationalist government not only set up organs in the United States, such as the UTC and the NRC-NY's Technical Office, but they also worked with the less explicitly partisan CIE, the Chinese American and nonpartisan China Institute, and the ITA, which had originated as a U.S. government organ but had spun off into an independent nonprofit by the time it undertook to partner on China's large-scale FEA/MEA training program in 1945. In addition, a number of key individual Chinese technical experts spent long periods in the United States during the war both advocating for and facilitating training opportunities. A deeply interconnected support network gradually formed over the course of the war and continued to facilitate training of Chinese students and technicians well into the postwar period.

The people who comprised this network were diverse in origin, experience, and perspective. In addition to American educators, industrialists, and government bureaucrats, the network included both Chinese-born and overseas Chinese and Chinese trained in both China and abroad. Not all Chinese members of this network were politically aligned with the KMT, but they were all willing, even eager, to work with the Nationalist government to defend China and plan for China's future, and they associated themselves with groups and institutions that had been constructed by the Nationalist government such as the UTC and the NRC. Important to note, however, is that even though they worked together to reach a set of common goals, they did not all agree all the time about how the Nationalist government should plan and do things. They made these debates public in publications such as the *National Reconstruction Journal*, which although designed to foster a sense of community and patriotism among trainees and students in the United States, also served as a forum for open and honest debate on policy and action.

Numerous different Chinese government entities attempted to utilize overseas training to address specific knowledge gaps that they needed to fill in order to achieve specific goals. We have seen in the examples presented in this chapter attempts to use overseas training programs to develop expertise to solve problems in such diverse fields as oil field development, aircraft engine manufacturing, and agricultural extension and mechanization. Even though such training programs required a significant time investment, both the trainees and their home institutions were willing to make that investment in the expectation that it would bring results and have an enduring impact. In the case of the thirty-one, many of the returning trainees went back to work on projects at the Yumen Oil Field in Gansu, where they improved oil extraction, storage, processing, and shipping methods. In the case of the aeronautical engineers, not only did returnees set up China's first aircraft engine plant, but they also established their own training program at the plant. In both of these cases, the trainees have been widely credited as being the fathers of their fields in China.

Finally, it is also important to note that these were ongoing activities that did not simply stop as the war ended. Although many were planned and understood as training activities that had to do with the war, others were initiated with postwar reconstruction in mind. All of these activities, however, depended on the goodwill of Chinese, Chinese American, and American allies, goodwill that in many cases was generated by a sense of having a shared cause against a common enemy but that was also not devoid of self-interest, particularly in the case of U.S. companies. Both Chinese and American actors were seeking to position themselves to advance their own interests in the postwar world. The Nationalist government, actively planning, as we have seen in chapter 1, for postwar redevelopment of China, was eager to build a cadre of engineers and skilled technicians who would be able to lead that development and to utilize any avenues the United States would provide to enable them to do so. On the American side, U.S. businesses were especially interested in developing markets for their goods, but during the war, in some cases, they were eager to take Chinese trainees simply because they were a source of reasonably well-educated personnel, which was in increasingly short supply in the United States as the war dragged on. The U.S. government was motivated by a desire to strengthen its ally in hopes that it could hold itself together against Japanese offense, continue to occupy Japanese military attention, and perhaps even take on a greater role in the fight against Japan. Chapter 4 will explore other ways in which the U.S. government sought to build China's industrial and agricultural capacity during the war.

CHAPTER 4

China's War Production Board

Technical Collaboration for Wartime Purposes

Grappling with the dual goals of making agriculture and industry immediately productive for war purposes and preparing western China to integrate more fully into a postwar national economy required more than just plans and trained workers. It also required institutions that could facilitate the achievement of these goals. Over the course of the war, the Republic of China (ROC) government reorganized its existing institutions and their duties in an effort to create a more streamlined structure that would be better equipped to oversee the kinds of development that the government wished to promote. These efforts, however, were not entirely successful.

The Nanjing government had created a number of seemingly competing institutions such as the National Economic Council and the National Resources Commission. However, by the late 1930s, not all of these institutions continued to exist, and in a series of reorganizations in the late 1930s and early 1940s, authority over both industry and natural resources came increasingly under the domain of the Ministry of Economic Affairs (MEA), under which the National Resources Commission (NRC) was subsumed. At the same time, however, the war and the consequent move to the interior also gave rise to a number of new institutions that oversaw the development or distribution of specific resources. By the latter stages of 1944, Weng Wenhao and the MEA appeared to be largely in command of many of the institutions most crucially connected to industrial development but still needed to cooperate with other ministries such as the Ministry of Communications, Ministry of War, Ministry of Agriculture and Forestry (MOAF), and Ministry of Finance in order to achieve development goals. Government decision making was not streamlined, and even though there were planning processes (discussed in chapter 1) intended to unify government action, government agencies were sometimes at odds with each other because their priorities differed, their leaders did not see eye to eye, or they failed to recognize each other's authority. Moreover, the government did not command all industry in inland China. In some sectors, the state was able to wield considerable authority. For example, as Ying Jia Tan has shown for the power sector, the Japanese invasion broke down earlier resistance to nationalization as it became increasingly evident that the state would need to lead efforts to develop a power infrastructure in the inland.[1] As Kwan Man Bun has shown in the case of Yongli Chemical Industries, however, some privately owned companies that migrated west and worked to develop new plants in unoccupied areas had complicated relationships with the state, seeking government support for their critical war work manufacturing chemicals but successfully repelling government attempts to bring them under direct state control.[2] Many other factories were small and independently owned, often erected rapidly by small-scale entrepreneurs seeking to profit in the rapidly and constantly shifting wartime economic environment. Although these private factories were affected in various ways by government policies, they were not in any direct or indirect way supervised by any government agency. Industry in western China was thus variously under the direct control of the state or privately owned, meaning that the state did not have comprehensive control over the use of resources or the production of goods. In addition, the circumstances in which both state-owned and private industry operated were extraordinarily challenging. In particular, the transportation, communications, and power infrastructure was underdeveloped and industrial materials and fuel were often not available. As a result, production of essential goods, including those needed to support the war effort, was uneven, unpredictable, and unreliable.

The Creation of the Chinese War Production Board

Into this situation walked Donald M. Nelson, architect and leader of the U.S. War Production Board, who was sent by President Franklin D. Roosevelt on a mission to survey China's war industries with an eye to helping China become less dependent on American materials and finished goods, all of which had to be flown over the hump from India at great expense. The timing of the Nelson mission coincided with a ramping up of U.S. military aid and loans to China, and although the ostensible rationale for the Nelson mission was to aid China, it was clearly the case that Nelson was also there to push China to maximize and streamline the support that it could provide to joint military operations against Japan. Nelson's focus, therefore, was on thinking about how to make Chinese industry more efficient and productive while working as much as possible with locally obtainable materials. As Nelson reported to President Roosevelt in December 1944, he had been dismayed to find that "Chinese arsenals have been operating in the midst of war on only 55% of capacity. Operating rates of most other industries have been even lower. The steel industry has been operating at less than 20% of capacity."[3] In his view, the main obstacles to productivity were structural. From his perspective, institutional problems were largely to blame for declining productive capacity in free China in 1943 and 1944, declines that had happened even in spite of the fact that the number of factories of all sorts in the region had increased rapidly in the period from 1938 to 1943, as had the range of agricultural activities necessary to support increased industrial production.

Foreign observers consistently criticized Chinese industry for failing to produce and for being overly dependent on Allied aid. This is a constant refrain in comments on inland Chinese industries written by both U.S. and British observers. Nelson, for example, was quite frank in September 1944 conversations with both Chiang Kaishek and Weng Wenhao about how unreasonable it was for China to request that the United States fly in heavy items like iron and steel and munitions when its own industry was not operating at full capacity.[4] However, few foreign observers appear to have grasped the range of challenges that ROC industries faced. Even in the case of well-equipped factories, it was unaffordable for industries to run at full capacity owing to the high costs or limited availability of raw materials, including fuel/energy, and high costs of transportation. Therefore, even though there is ample evidence that the Nationalist government was working to increase production in China of items such as munitions, vehicles, fuels, textiles, and food that were necessary to both the conduct of the war and the rebuilding of China after the war, many of China's American advisers seemed to think that the fundamental problem was poor leadership.

Nelson also felt that both production and efficient use of resources was being hampered by a disconnect between wartime planning and post-war planning that he felt pervaded Chinese government ministries, a disconnect that was most prominently evident in the Nationalist government's failure to swiftly utilize all of its manufactures and other resources and to, instead, safeguard some of them for postwar use. "The various Chinese ministries have the notion that wartime economy and post-war economy are entirely two different things and that conditions existing during the war can be completely changed in the post-war period." Nelson observed to Chiang, "I wish to remind Your Excellency that there will be a way out after the war only when there is a way out during the war."[5] In other words, what China did during the war would put it on whatever path it would be on after the war. It could not, or at least should not, harbor resources during the war in hopes that it might put them to better use following the war because to do so might mean that China would lose the war. Nelson's concerns were entirely focused on the crisis at hand.

Nelson's perspective on the root cause of China's industrial production woes was that leadership was failing and China lacked an institution that could guide the change that was needed. Not one to look a gift horse in the mouth, Weng Wenhao asked him to provide American advisers to come guide industry toward higher production. Nelson replied that he would be happy to help but that he wanted first to see a new Chinese structure put in place to oversee production, one that had a single person in charge who had the ability to make decisions. China, he said, "is getting a poor reputation in the U.S. Our people return to the states discouraged because necessary things don't get done here." When Weng offered the opinion that a committee could be set up, Nelson responded that he did not like committees, which were "inclined to only talk." Instead, he wanted to see China put into place a structure where "one man [would] make the decision, receiving advice first from others. . . . You need to set up a War Production Board here and demonstrate ability and willingness," he observed. "I don't see any use of planning a post-war program when only a few simple government difficulties can now hold up production." Nelson's approach to rectifying this situation was to propose that the Chinese government re-create an American institution in China with the expectation that by streamlining the decision-making process, eliminating decision making by consensus, and placing authority in the hands of a single decision maker, Chinese production would be improved and China would better position itself for postwar redevelopment. "You need a proper organization to work out your war problems and out of that experience will grow the know-how for doing the post-war problem," Nelson told Weng.[6]

Nelson's high expectations of the role that could be played by a single well-designed institution came directly from his own experience running such an institution in the United States for the previous couple of years. The American War Production Board, which had been created by Roosevelt in January 1942, just weeks after the United States entered World War II, was designed precisely this way. As Nelson later wrote of the creation of the American War Production Board, Roosevelt had selected a single man, Nelson, "who could bring together all discordant elements—whether through cajolery and compromise or, at last, by plain brute force—and who could get industrial production organized and going in a volume never dreamed of before, and get it going quicker than instantly."[7] Roosevelt had endowed Nelson with considerable authority and had set him loose on industry, agriculture, general economic policy, and the military, asking that he mobilize America's "industrial strength for the manufacture of the weapons, supplies, and accessories of modern war."[8] As chairman of the American War Production Board, Nelson answered directly to the president and was empowered by executive order to serve as director of the war production and procurement program and to determine "policies, plans, procedures, and methods" of government agencies with respect "to war procurement and production, including purchasing, contracting, specifications, and construction; and including conversion, requisitioning, plant expansion, and the financing thereof; and issue such directives in respect thereto as he may deem necessary or appropriate." The order further decreed that the chairman's "decisions shall be final."[9]

Nelson expected that China's War Production Board would be granted similar authority so that it would be able to direct the activities of both government ministries and private enterprises engaged in war production work. Unless authority was invested in a single institution that could develop overarching policies to guide, restrict, and/or stimulate both consumption and production, Nelson figured, the Chinese government and the economy it oversaw would continue to fail. The Chinese War Production Board (CWPB) needed to be empowered with the strength to get things done, which meant that it should answer directly to (and only to) the generalissimo. It should not, Nelson observed, directly oversee any government agencies or take it upon itself to carry out the things that needed to be done. Rather, it should make decisions that would guide the actions of the appropriate ministries, agencies, and enterprises. In addition to "directing, supervising and coordinating the various public and private production agencies," the CWPB should, in Nelson's view, make decisions related to transportation, allocation of materials, standardization of techniques and production, and commandeering and conservation of materials.[10] If the Chinese did not create a War Production Board, Nelson made clear, they could not expect much in the way of U.S. assistance with either wartime or postwar problems.

Chiang Kaishek, Weng Wenhao, and the Chinese government were in no position to resist Nelson's heavy-handed approach. They were hamstrung by their need for U.S. loans, technology, and advice that they believed would serve them in both the short and long run. On November 16, 1944, the date of Nelson's return for a second visit to China, the Chinese government announced the creation of its own War Production Board (戰時生產局, CWPB) that was closely modeled on the American War Production Board. A month later, it was officially inaugurated. With this move, Nelson was satisfied that U.S. methods and U.S. know-how would provide the answers to Chinese industrial-military problems.

Like the American War Production Board, the primary aim of the CWPB was to strengthen war production in China. In particular, it focused on increasing production of iron, steel, and nonferrous metals, ramping up production of coal and coke to support steel manufacture, developing electric power in the region's major cities, increasing fuel production to facilitate military activities as well as transportation of raw materials and manufactured goods, and increasing production of military equipment. By way of oversight, the CWPB placed orders for supplies, stockpiled products so as to be able to distribute them, leased equipment and made loans to factories to help them improve capacity and production, offered technical advice to improve efficiency, and established standards. In addition, the CWPB also served as a clearinghouse for all lend-lease applications to the United States except for those related directly to military matériel and transportation and handled purchasing of all items abroad for both public and private institutions.[11] Its organic law, passed on December 6, 1944, stipulated that the CWPB would "exercise complete authority over the work of war production" and that its aim was "to obtain maximum production for war and essential civilian purposes, and to direct, supervise and coordinate the war production organizations, both public and private." It was granted a broad authority to set production priorities, allocate materials, facilities, and personnel, import and export war materials and civilian supplies and prioritize transportation of goods, set standards, stockpile, and set limitations on various kinds of work.[12]

Organizationally, the CWPB was headed by Weng Wenhao, who was advised by an advisory committee made up of a group of Chinese area experts whom Weng had selected and a small group of American technical experts, most of whom had prior experience working for the American War Production Board. These men comprised the American War Production Mission in China (AWPM). Under Weng were a set of ten departments that were organized not by industry but rather by bureaucratic activity (secretariat, purchase, personnel, etc.) and a number of industrial advisory committees (liquid fuels, iron and steel, coal and coke, auto parts, etc.) that met regularly, included representatives of both state-run and private industry as well as American subject experts, and advised Weng on policies and procedures that they believed would improve production in their particular industry. In addition, the CWPB opened branch offices in Kunming and Lanzhou from which locations it could better communicate with industries and bureaucrats in the southwest and northwest.

Weng recruited his staff in late November and December 1944 but was in constant contact with AWPM leaders as he did so. In a number of cases, he was urged by AWPM staff to recruit Chinese then residing in the United States, such as T. P. Hou, to return to work for the CWPB, and records show that he did attempt to do so, but it is unclear whether he succeeded in many such cases.[13] By April 30, 1945, Howard Coonley, outgoing head of the AWPM who had served as Nelson's deputy in China, was reporting to Nelson that the CWPB had a staff of about five hundred and that it was operating along lines that were similar to the American War Production Board.[14] Recognizing the importance of making the institution run as effectively as possible, Weng and Coonley agreed that it would be important to bring specialists to China who had direct experience with the operations of the American War Production Board and its associated industries.[15] The activities sponsored by the CWPB were financed by a ten-billion (Chinese dollars) line of credit extended by China's four major banks.[16]

China wanted similar results to what Nelson was looking for, but Chinese leaders consistently argued that to achieve such results would require a well-developed program of technical assistance. In his negotiations with ROC leaders including T. V. Soong, Weng Wenhao, and Chiang Kaishek in September 1944, Nelson was repeatedly asked if the United States could provide "technical men" to help get various industries, particularly steel and munitions, on track. Chiang Kaishek made it clear that what he wanted was real help and not just observation. "The men sent to China by the American Government in the past," he said, "whether military or political personnel, never proposed any ways and means of improving Chinese products and increasing Chinese production."[17] Nelson promised to send experts on textiles, steel, power, consumer goods, export trade, railroads, and banking, though by the time of his second visit to China in November 1944, the group he had rounded up to take back with him included only five experts on steel production and one expert on alcohol production who were to study Chinese production methods, and work with plant managers and government officials to increase output, improve quality, and reduce costs.[18]

The American War Production Mission

Nelson created the AWPM to both keep his promise to provide technical advisers and keep men on the ground in China who could monitor the activities of the CWPB. The AWPM included men from U.S. industry and U.S. government, most of whom had specific technical expertise in areas of concern to the CWPB and many of whom had already served in the American War Production Board. Leading the mission, though only sometimes in China, was Nelson, who was replaced in the summer of 1945 by Edwin A. Locke Jr., his right-hand man on the American War

Production Board. Both Nelson and Locke conducted much of their work on the project from the White House. Running the mission in China were Deputies in Charge of Mission Howard Coonley, James Jacobson, and Andrew Kearney, all of whom had worked for the American War Production Board since 1942. Other regular mission staff included special assistant Edward Candee, administrative assistant Francis Cleary, Lester Bosch, who was in charge of production specialists, and Rudolph O. Johnson, who was in charge of reporting on production. In addition, the AWPM brought in numerous industrial consultants to work on specific problems. The work of finding such experts was undertaken as a special project by White House staff. By the beginning of 1945, the White House was regularly hunting down appropriate American experts and sending them to China to work for the AWPM. For example, when the CWPB indicated a need for someone with expertise in coal washing (to take out impurities and improve quality), it was a technical adviser to the personal representative of the president who started calling around and writing letters to industry representatives as well as government men to identify an appropriate technician.[19] The fact that such requests were coming from the head of the American War Production Board, which had been invested with great authority over both public and private enterprise in the United States, almost certainly made both groups more responsive.

Most of the AWPM staff were recruited from known companies and government agencies that were already familiar to the AWPM leadership, but the AWPM's personnel files clearly show that the project was of some interest to American technicians wishing to get ahead in their careers. One twenty-two-year-old metallurgist who had graduated from the University of Illinois wrote in, saying that he had read about the program in his local paper and "it occurred to me that this mission might have use for the services of one with a knowledge of chemistry and metallurgy." Louis Anderson already had a job, but he noted his draft status in his letter, so perhaps he was looking for a way to serve that would not involve fighting. Although his motivation for wishing to head off to China as part of the AWPM was unclear, he was eager enough to do it that he had taken the initiative to send in a letter of inquiry.[20] The AWPM did take some of the many volunteers who wrote in seriously enough to go to the trouble of doing background checks with the Federal Bureau of Investigation and getting references from their work colleagues.[21]

By the end of the mission in December 1945, fifty-three different men had served in the AWPM, forty-one of whom were industrial or production consultants, more than twenty-six of whom had been recruited directly from industry, and others of whom had been recruited from U.S. government organs. Of those who had been employed by the War Production Board in the United States, a number left their prior positions in industry to take up those posts, but some continued to wear two hats. Howard Coonley, for example, served as chief of the simplification branch of the conservation division of the War Production Board from March 1942 to November 1944 but concurrently continued in his role as chairman of the board of the Walworth Company, where he had worked since 1913. Tom Kearney continued on as a partner in the industrial consulting firm of McKinsey, Kearney & Co. Thus, both of these men had clear and direct ties to industry even while they were serving on the AWPM. Coonley and Kearney both held leadership positions and remained in China for longish periods, but the AWPM also hired short-term consultants to spend just one or two months in China. Their salaries were paid by the Foreign Economic Administration (FEA), who paid them at their current rate plus an overseas allowance and travel expenses, and in many cases, their actual room and board in China was paid for by the Chinese government. Among these men were quite a few who were granted temporary leave from large companies like Bethlehem Steel and the Pittsburgh Coal Company, medium-sized companies like the McLean Iron Works and Koppers Construction Company, and small independent consulting firms.[22]

A major reason that companies were willing to part with personnel for a time was the hope that those personnel might pave the way for future industrial opportunities in China. For example, LeRoy Whitney, whose job involved recruitment of industrial consultants, noted in a memo to his boss that Blakeslee Barnes, vice-president of the American Cyanimid Corporation, had agreed to make one of his employees available for a fee. Barnes had told Whitney that they would train this fellow up on a low-temperature coal carbonization process his company had just developed, "all of the equipment for which Mr. Barnes believes can be built right there in Chungking." Whitney observed that "if this plant will do what Mr. Barnes claims, it will solve the coke, gas and by-product problems in Free China."[23] It does not appear from this description that Barnes was looking to sell machinery to China at that time, and in fact, the AWPM was only in China to advise, not to sell. However, it does seem that had his company's method been adopted in China, that might have created opportunities for future engagement and possibly for future sales or at least consulting fees. The potential for profit surely motivated corporate willingness to provide personnel for the program.

Certainly, at least in some industries, American businesses were already being primed by 1944 to think about doing postwar business in China. Edward G. Whittaker of the China-American Council of Commerce and Industry, Inc., for example, promoted this idea for the cotton industry. An article titled "China: Huge Potential Market for Textiles and Textile Machinery?," which was based on an interview with Whittaker, argued that "to replace its stripped and sabotaged textile industry, China will need new equipment at the rate of 500,000 or 600,000 spindles and 20,000 looms a year after the war. Her needs for raw cotton and cotton cloth will be enormous, and we can help supply them." Postwar China was being touted as a vast new market for surplus American products "when the war machine will no longer be absorbing half the national product."[24] If U.S. businesses were concerned that the end of the war would mean declining revenue, China and other places like it could fill the gap and replace the U.S. government as a major consumer of American products. AWPM consultants would almost certainly have brought similar assumptions to their work in China.

The AWPM's industrial consultants spent a good deal of time in the field, making firsthand observations of China's industries and reporting back on their findings to the AWPM and the CWPB. Their aim was to acquire "a thorough knowledge of the problems of the producing units in each major industry" and to plan "the specific steps required to increase production, to lower costs, and to improve quality." Each consultant developed such plans independently, their plans were merged into an overall production plan for the industry in question, and those plans were "reviewed in conference with officials of the War Production Board, and revised in accordance with their views." In other words, although the American industrial consultants had considerable input, their recommendations were not always taken wholesale.[25] Nonetheless, the staff of the AWPM reported in late May 1945 that AWPM specialists were doing too much of the work in large part because the CWPB was finding it hard to identify Chinese technicians with the experience and abilities needed to serve as strong deputies to Weng Wenhao and similarly hard to attract such men into the service of the CWPB even when they did find them.[26]

Goals of the CWPB and AWPM

AWPM technical consultants brought with them a set of American assumptions and expectations, and although many of them adapted their views as they became increasingly familiar with the industrial environment in inland China, the overall tenor of the mission never ceased to bear the imprint of Nelson's War Production Board approach. Various sources ranging from reports on industrial advisory committee meetings to suggested commendations for special service to the CWPB show that AWPM consultants were most concerned by (1) the need for better coordination, (2) lack of standardization, and (3) diffuseness of production. In addition, as we shall see later in this chapter, the consultants were also well aware that poor transportation and infrastructure further exacerbated efficiency problems in a variety of ways.

Although some of these challenges, such as standardization and the development of transportation infrastructure, were issues that the Chinese government had been working on prior to the war, some of them resulted directly from the wartime circumstances in which the Chinese government had to operate. During the early war years, there had been a scramble by both the state and private entrepreneurs to construct new industries and increase production. Nationalist government agencies, concerned in particular about keeping new plants safe from Japanese bombs, often located them in out-of-the-way spots, including caves, as we have seen in chapter 3. These safety concerns often trumped access to raw materials or transportation infrastructure when new plants were set up. Many of these problems might have been solved with the rapid development of a better network of modern roads and improvement of the fuel supply, but although resources were poured into road construction during the war, roads were not modernized at a fast-enough pace to enable industry to operate smoothly. Private entrepreneurs were guided by somewhat different concerns and tended to use whatever capital they had to set up often very small factories wherever they landed. Because they were driven first and foremost by the desire to earn a profit, they typically built their factories in close proximity to raw materials, but that did not necessarily mean they had good sources of power or even transportation, and sometimes state restrictions on the use of raw materials created impediments to production. Private enterprises were also less likely to produce goods that met whatever standards state agencies agreed were necessary for any given industry and the state's capacity to control such enterprises was limited.

Both the CWPB and the AWPM sought to rectify these organizational and infrastructural problems, at least for China's most critical industries, but they were not always entirely in sync in the way they approached this task. In particular, members of these groups differed on the question of just how fully China could adhere to an American model. Nelson was looking for China to adopt wholesale an institution and a way of doing things that had been shaped by specifically American experiences and conditions. CWPB members were operating in a wholly different environment. They were embedded in political and infrastructural systems that had been shaped by specifically Chinese experiences and conditions. For as much as both American and Chinese members of these organizations may have wished to import American institutions and ways to China, those institutions and ways would have to interact with a new environment and would inevitably cease to be recognizably American in that process.

NEED FOR BETTER COORDINATION

One issue that was quite apparent to both AWPM and CWPB staff was the problem of coordination between the various entities in the Chinese government on the one hand and between those institutions and U.S. agencies in Chongqing on the other hand. Walter G. Whitman of the American War Production Board, writing to James Jacobson of the AWPM, quoted an anonymous "Chinese of considerable technical education and capacity as having written to him that 'The problem on hand is how could we best utilize the *limited* facilities to make the largest quantities of most useful products out of the available raw materials.' The problem itself sounds natural and simple enough. But how to solve it is another matter. It seems to me that the industries in China have been mostly handicapped by lack of coordination and lack of efficiency. . . . The proposed central organization [CWPB] must have immense power to bring together and control all the agencies running China's industries now. It's bound to meet head-on collisions. But until it is able to disregard the objections there really is not much that can be achieved. The lack of efficiency is not only apparent in the interdepartmental matters within a factory, it is even more obvious when a factory deals with other organizations. In other words, it is one of the consequences of lack of coordination."[27] This perspective dovetailed with Nelson's impression, guided in large part by his own past experience with the American War Production Board, that what China needed was a strong, central organ that could bring all parties together to achieve goals of common interest. To Nelson, the establishment and maintenance of a positive spirit of cooperation between all of these groups was a fundamental goal of the AWPM and the CWPB.[28]

Of particular concern to the AWPM was "that working arrangements and understandings be had with other agencies of the Chinese Government." To facilitate this, some organs, such as the Liquid Fuel Control Commission and the Industrial and Mining Adjustment Administration (previously under the NRC) were brought under the direct authority of the CWPB. Cooperation agreements were reached between the CWPB and other units such as the Chinese army's ordnance department, the Chinese War Transportation Board, and the Chinese office of the United Nations Relief and Rehabilitation Administration.[29] Coordination with other government organs was challenging, however, as the heads of various ministries were not entirely willing to recognize the CWPB's authority. Moreover, in spite of the fact that the CWPB had been endowed with full authority over production, to exercise this authority it needed the financial capacity to stimulate production of certain items through procurement, and the Executive Yuan refused to provide it with the funds that it would need for such action.[30]

The seven committees of the CWPB, each of which focused on a different industry, were designed to create precisely the kind of coordination that China seemed to need. Each committee consisted of representatives of the relevant Chinese government organs and NRC industries, as well as the relevant technical advisers from the AWPM. The goal of each committee was to identify ways in which the industry as a whole could be made more productive, which sometimes meant identifying bureaucratic obstacles to productivity. AWPM and CWPB staff did not, however, see eye to eye on all points. For example, whereas AWPM staff wished to see representatives of private enterprise well represented on the committees, at least in part because of the perspective they might bring to the problems under discussion, the CWPB leadership showed little enthusiasm for engaging with private companies. Weng Wenhao responded to a suggestion that the CWPB engage more fully with private enterprise by extending manufacturing contracts to small plants somewhat dismissively, telling Howard Coonley that most private enterprises were too small and "poorly equipped" to be able to produce the kinds of materials that the Chinese government required.[31] This was in fact true, though it was also true that some private firms operated in competition with state-owned enterprises. The American advisers' insistence on the greater involvement of private enterprise was based on their knowledge of American industry and their recent experiences with the U.S. War Production Board in combination with a limited understanding of the Chinese industrial landscape. Chinese industry, particularly in western China, looked very different from industry in the United States. The NRC managed the bulk of China's most successful heavy industries. Most private enterprises in western China operated on a very small scale and for Weng and CWPB leaders, there was simply no logic in trying to engage with those enterprises. The AWPM, like the U.S. War Production Board, saw private enterprise as an important partner with government and its advisers expected that a truly functional coordinating entity would facilitate and direct that partnership.

Coordination of effort across industries was another issue that AWPM advisers felt needed to be overcome. As Calvin Joyner, FEA representative in Chongqing, observed to Nelson, Locke, and Jacobson on the occasion of their first visit to Chongqing, there was too much duplication of effort. As he put it, "each plant makes its own bricks, each has its own vegetable gardens, each department makes its own tools."[32] If plants could be discouraged from diversifying to this extent and instead encouraged to focus all their efforts on their area of specialization, that would improve productivity. This concern, while valid, did not adequately take into account the real situations of Chinese factories. In fact, Joyner understood that the root cause of this diversification was poor transportation that affected the movement of both raw materials and manufactures. Even though he surely also understood that the plants had no other choice than to devise their own strategies to meet their immediate needs, he still wanted to apply an American organizational approach to the Chinese case. This way of thinking, a sort of befuddlement at the way Chinese industries did things, was not uncommon for American observers, particularly those who did not spend much time conducting site visits to actual factories.

The CWPB faced numerous obstacles as it sought to play its coordinating role. One of those was simply the challenge of managing different government organs that did not all relate to each other effectively. However, other situational problems, caused by the ways in which both private and state-owned enterprise had developed in the region, also contributed to the lack of coordination, and these problems were seemingly just as intractable. In spite of ample evidence to the contrary, it appears not to have been immediately obvious to either Nelson or some of his AWPM advisers that American-style efficiency might be hard to achieve in wartime western China.

QUEST FOR STANDARDIZATION

Efficiency was clearly something the AWPM consultants admired and hoped to encourage because with efficiency came productivity. When they saw it, they noted it and thought it worthy of commendation. In recommendations for special merit awards that AWPM consultants submitted to the CWPB in the fall of 1945, the consultants consistently recognized factory managers for their efficiency. Lionel Booth, a metallurgist consulting on the concentration and dressing of ores, who was in China from mid-May to early September 1945, praised Sze Chia-Foh, vice-director of Kunming Electro-Metallurgic Works, for operating the plant "in a very business-like and efficient manner. One could not help but notice that Mr. Sze made his plans well ahead of schedule, his operations were never without the necessary products to carry on the work, such as fuels, fluxes, and various chemicals required in the plant work."[33] Henrik Oveson, specialist in iron and steel plant engineering, who was in China from mid-November 1944 to late February 1945, wrote that the Ming Sung Ship Building Co., one of the region's few large private enterprises, should be commended "for building an efficient shipyard with shops under most strenuous conditions."[34] Ralph Strang, specialist in machine tools who spent March through November 1945 with the mission, noted the efficiency of the 53rd Arsenal in Kunming and of the Dai Chuan Industrial Company where "every department is run most efficiently and production is high."[35] Perhaps one reason that American advisers were so impressed by efficiency was that inefficiency appears to have been a much more common occurrence. Unused industrial capacity and lack of materials that would permit factories to produce at capacity were among the most common issues that AWPM advisers noted, but among the manifold causes of inefficiency, of particular concern to the AWPM was the apparent absence of universal industrial standards in China. To that end, standards became a point of emphasis for regular mission staff and specialist consultants alike.

The lack of standards appeared to mission participants to be a major obstacle to production. As Howard Coonley wrote to Weng Wenhao in January 1945, "It is becoming evident that more and more importance will be attached to the development of Standards including material, design and quality specifications for the products being contracted for by WPB." Some standards, he told Weng, could be taken directly from the United States, but there were other things for which the United States also lacked uniform standards, and for such cases, Coonley felt, the CWPB should "set up committees to develop War Emergency Standards" and appoint Dickson Reck, an American consultant on standards who had gone to China under the cultural relations program, as a standards liaison.[36] This issue continued to concern Coonley even after he had left the AWPM. As he wrote to Whiting Willauer of the FEA in April 1945, "In developing a program of increased production of war supplies in China, the War Production Board of the National Government of China is faced with a serious difficulty. Due to the lack of specifications suited to the materials and equipment that are available and to shortage of testing instruments and chemicals, they are facing serious problems." The problem that Coonley identified was not so much a complete absence of standards but a lack of specific standards for the particular types of materials and equipment that Chinese industries had to work with. To address this problem, Coonley wrote, "The Chinese WPB has set up some twelve technical committees to create emergency standards in line with those established by the American WPB in cooperation with the American Standards

Association." In addition, the CWPB appointed the National Bureau of Industrial Research (NBIR) to undertake testing of products and equipment to ensure that they matched the new standards.[37]

The technical committees, which worked on a variety of problems in addition to the standards, eagerly attacked this role. One problem that some of these committees identified was that factories that had sprung up quickly, often operated with homemade or otherwise imperfect machinery, and frequently had to manufacture items using unusual materials, were not always able to produce the kinds of high-quality and standardized products that would be necessary to facilitate efficient interchange of parts across factories and industries. For example, according to Ralph Strang, when the Machine Industry Advisory Committee discussed standards for lathes, the U.S. technicians on the committee "suggested that the number of lathe manufacturers be reduced" because the shops they had visited were not making good lathes. Part of the problem the committee identified was that the lathe-making machines were not standardized. "Mr. Brown made the suggestion that [after some manufacturers were shut down] these remaining manufacturers break down the manufacturing still further and that each one of them concentrate on some parts of the machine and then have the lathe assembled from standardized units. This certainly will increase accuracy, efficiency and production, and, at the same time, greatly reduce costs."[38] Both Brown and Strang, the AWPM representatives on the committee at that time, were of the opinion that industrial production would work better if all of the producers of related goods could work together and operate from the same standards.

Similarly, the CWPB's Machine Industry Advisory Committee was especially concerned about standards for gauges in machine shops that produced machinery needed to make ordnance. They proposed the establishment of "a central checking place . . . as soon as possible to provide both the user and the manufacturer of gauges a place where he or they could bring the various gauges for an accurate check. This would help greatly toward the establishment of standard products conforming to uniform specifications, and also providing a place where periodic checks could be made to standards enabling the various manufacturers to maintain their gauges in the best condition possible." To address this problem, the committee proposed to set up control bureaus in Chongqing and

Kunming that "would make available precision testing equipment to even the smallest manufacture [*sic*] who could not otherwise obtain the benefit of such equipment."[39] Such moves might have been useful for plants located near these major centers but perhaps not for the many plants that were spread out in other parts of inland China. Geography would surely hamper such strategies to coordinate, centralize, and standardize.

The Americans were not the only ones concerned with standards. The Nationalist government, in the 1930s, had already set up some institutions that were supposed to be overseeing at least some standards for industry and agriculture and the NRC had initiated preliminary efforts to standardize certain industries, such as power, in the prewar and early war periods.[40] By and large, however, most industries had not yet been standardized. Among the institutions designed to identify and promote standards was the NBIR (中央工業試驗所). Running in a distant second place but performing tasks similar to the NBIR was the Ministry of Forestry and Agriculture's National Agricultural Research Bureau (NARB, 中央農業試驗所). Both of these institutions had been created in the early 1930s, but their roles became clearer and more important during the war as they took responsibility for establishment and maintenance of standards, development of knowledge and tools that could be utilized by farmers and industry, and extension of that knowledge to their partners in the private sector.

The NBIR had been established in 1931 but was reorganized in 1938 following the westward retreat. Headquartered in Chongqing, by the latter part of the war it had sixteen laboratories and twelve experimental plants. Many of these were located in Panchi, just outside of Chongqing and directly across the river from National Central University, but some had been placed in strategic locations that were particularly close to raw materials. For example, sugar research and experimentation were conducted in Sichuan's Neijiang, a major sugar-growing region, and the salt laboratory was near the salt well at Tsilutsing. With time, the NBIR also opened branches in other parts of western China, such as a northwest branch office in April 1944. By locating branches across the region near important industrial sites, the NBIR positioned itself to disseminate centrally determined standards, advice, and mandates as well as to collect information about industry and how it functioned that could be funneled back to the center. In these ways, it served the needs of both industry and the state.

An important role of the NBIR was to facilitate both modernization and standardization. As Y. T. Ku wrote in a report on the activities of the NBIR, "The manufacturing technique and process in our industries need improvement and modernization," and the NBIR's aim was to assist in that process.[41] The primary role of the NBIR was to facilitate industrial development by conducting research on industrial raw materials, techniques, and processes and conveying the results of that research to industry. In some industrial sectors, scarcity of typical raw materials and industrial machinery had led both state-run and private enterprises to experiment with a wide array of alternative materials and jury-rigged machinery to make items that were in high demand. Variations in the quality and nature of raw materials and in the quality and suitability of the industrial process led inevitably to variations in finished products. Although such variations were not necessarily significant in all types of industry, machine and automobile parts as well as ordnance were all products that absolutely had to be manufactured according to particular specifications in order to be usable. Better knowledge of the properties of the materials at hand, better understanding of the most efficient ways of manipulating those materials, and a better grasp of appropriate industrial processes for working with such materials were all goals that the various laboratories and experimental plants of the NBIR aimed to achieve.

Deeper knowledge of materials and processes would be most meaningful in combination with a uniform set of standards. To address that need, by early 1945, the NBIR had been asked by the newly established CWPB to take on the role of developing industrial standards and inspecting and testing industrial products to ensure that they complied with those standards. The aim of these activities was to elevate overall industrial output of usable products. AWPM advisers had identified the NBIR as an appropriate institution to undertake at least some standardization work in January 1945, when E. K. Waldschmidt reported that the NBIR "represents the best facilities in Chungking for chemical and physical testing" and went on to "suggest that the War Production Board establish a Central Laboratory under the personal direction of Dr. Ku in order that the steel plants in this area may have some place at which they can get chemical and physical checks of their product."[42] This plan was consistent with Ku's own intentions for the institute. He was happy for the NBIR and its subsidiary units and model factories "to serve as a guide for the private enterprises as to the technical process, standard of product and method of management and in some cases training personnel for them." However, he also wanted factories to be able to conduct tests for themselves. As he wrote, "It is the hope of this Bureau to see that [a] controlling laboratory should be established in every large firm to test the materials and products. And to materialize this hope we made and supply to the firms all the important testing instrument, apparatus, chemical reagents, and also standard method of testing."[43] Although perhaps in conflict with the AWPM's desire to streamline the activities of Chinese factories, Ku's approach had the best chance of yielding consistent results across factories that were spread out over a vast region.

The NBIR was not the only standard-setting organ of the Nationalist government, although it was the largest and most important. The NARB played a minimal role with regard to industrialization and was thus not an institution on which the AWPM focused, but it, too, undertook to establish norms and standards as they related to agriculture. In the field of agriculture, standards were important both for understanding and describing the nature of agricultural output and also for the promotion of agricultural improvement. If farmers could be encouraged to plant seed that had been identified as maximally suitable for their region and land type and to utilize modern standardized planting, harvesting, and storage methods, then the overall agricultural output of the nation could be improved. The NARB was in charge of doing the underlying research to determine the norms that should be applied in any given region. Located, after 1940, in Beibei, Sichuan, but conducting research in experimental plots throughout western China, the NARB did work that was understood by the MOAF to be important to national economic interests and economic planning. Its main work was directed at identifying specific strains of rice, wheat, cotton, and some additional crops such as soybeans and potatoes that would grow well in the soils of southwest China and determining through experimentation what the optimal planting methods for those crops would be. In addition, the NARB established classifications and grades for these agricultural products. Finally, the NARB also produced reports on crop and livestock production in the provinces of western China.[44]

In addition to overarching organs like the NBIR and the NARB, there were also smaller, more focused entities that set standards for specific products, although not always entirely effectively. For example, standards for tung oil, which was of considerable national economic importance in the early part of the war, were set by the China Vegetable Oil Corporation (CVOC, 中國植物油料廠公司) using standards that had been developed by Chinese chemists in 1929 that were based on a set of U.S. standards established by the American Society for Testing Materials.[45] The CVOC, created in 1936 as a public-private joint venture, managed the refinement and export of tung oil and other vegetable oils that were produced by small-scale farmers, collected by itinerant oil peddlers, and eventually sold to CVOC collection stations. Although China had a virtual monopoly on tung oil production, CVOC leaders, who included both long-term government men and men with a background in private industry, nonetheless believed that the best strategy to protect and grow the Chinese tung oil business was to standardize the quality of the export product. To achieve this, before the war, CVOC factories set up modern oil presses and refining plants along with testing systems and high-quality storage tanks that would help ensure the purity of the oil they were marketing. Through application of standards that conformed to international expectations for tung oil, the CVOC was able to ensure that tung oil exports, in particular those exports that were used to secure a $25 million loan from the U.S. government early in 1939, would meet international standards. The raw tung oil the CVOC refined, however, varied in quality and purity, and the CVOC struggled to get local oil producers to adhere to these standards, thereby making their own refining and grading processes more complicated.[46] Although standards for raw oil had been generally agreed on well before the war, they were not universally applied. The Nationalist government's establishment of a tung oil monopoly in July 1939 did nothing to further these goals. The monopoly was necessary to ensure that the CVOC would be able to collect all the tung oil needed to repay the $25 million loan, which was funding Nationalist government purchases of nonmilitary supplies in the United States. But although the monopoly did work to ensure that tung oil was being put to the uses that government needed it for, over the course of the war it also led to a reduction in tung oil production, as farmers found that their profits were dwindling.

During the war, in part because exporting anything, including tung oil, from western China became increasingly difficult and in part because the government was trying to keep farmers producing oil, government scientists found new uses for tung oil. In particular, the NBIR and some NRC enterprises began experimenting with cracking tung oil for use as a biofuel in fuel-scarce western China. As a result, a whole new set of standards for cracked tung oil was developed by representatives from the CVOC, the Foo Shing Trading Corporation (a state-owned trading company that oversaw the tung oil monopoly), and the NRC's major tung oil cracking plants. By the time the CWPB's Liquid Fuels Advisory Committee took up the question of tung oil standards in January 1945, the standards for cracked tung oil had already been agreed on, but the group that had devised them remained concerned that their standards might not be universally implemented and was thus calling on the Liquid Fuels Control Commission (LFCC) to develop and apply a set of simple tests for all biofuel produced from tung oil to ensure that it met those specifications.[47]

Standards were an important concern of the AWPM advisers as well as the Chinese state-owned research units and manufacturers with which they worked. The AWPM encouraged the CWPB to set and disseminate standards that would assist in the industrial modernization process. The NBIR played a particularly large role in this regard, but it was not the only entity setting standards, nor was its capacity in terms of personnel or areas of expertise expansive enough to permit it to set standards for all areas. The NBIR was eager to take on this role, and other groups, including specific NRC factories and groups like the CVOC were also engaged in determining standards in their particular industries. As the case of tung oil suggests, however, even when there was a single entity, such as the CVOC, that had authority to regulate a particular product, producers could or would not necessarily follow those regulations. Moreover, even in cases where it was relatively easy to determine an appropriate set of standards, applying those standards could be extremely complicated. The tung oil case also shows us that, at least in some fields, the application of American standards to Chinese production was already being done well before the outbreak of the war. Chinese technicians and businessmen alike understood the importance of standardization. However, Chinese systems of production, particularly in industries that depended on small-scale producers, in combination with wartime challenges made it especially difficult to apply universal standards across an entire industry.

FIG. 4.1. Men at work at the cupola furnace at the Central Machine Works in Chengdu. Courtesy of Needham Research Institute.

DIFFUSENESS OF PRODUCTION

One problem that AWPM technicians and other foreign observers noticed nearly everywhere they went was that Chinese industry was fundamentally inefficient. This was largely a function of geography and of the state of disarray within which most factories in western China had been constructed. In general, both state-owned and private factories had been erected hastily and not always with regard to proximity to raw materials, fuel sources, or markets. Factories of strategic or military consequence, in particular, were frequently constructed in remote locations, often in caves like the Crow Cave airplane engine factory at Dading discussed in chapter 3, so as to keep them safe from Japanese bombs (see figs. 4.1 and 4.2). One downside of this strategy was that such factories were typically well off the main roads and far from the sources of raw materials, machine parts, and fuels to run machinery and thus dependent on trucks to bring those items in and roads to accommodate heavily laden trucks.

Poor roads and a scarcity of trucks isolated factories from mines and other raw materials in addition to distancing them from each other and

FIG. 4.2. Weapons being produced in an underground factory at the 53rd Arsenal near Kunming. Courtesy of Needham Research Institute.

from the markets for the goods they produced. As a consequence of isolation and scarcity of raw materials and power, factories often could not operate at full capacity. Reports on wartime industry by Chinese and Americans alike were replete with comments on factories operating well below capacity because they lacked either the raw materials from which their products were made or coal to fire up their furnaces. Edwin K. Smith, a metallurgist in China as part of the State Department's cultural relations program, wrote up one such report to Patrick Hurley in April 1945. On his visit to the NRC's Yunnan Iron and Steel Works, which had a theoretical capacity of forty-five tons per day and which supplied important NRC factories in Yunnan such as the Central Machine Works and the 21st Branch Arsenal, he found the foundry idle and was told that neither ore nor coke were reaching the plant owing to a lack of trucks to transport the materials.[48]

In addition, poor linkages with other factories that manufactured machine tools and parts, for example, led many plants to diversify their activities so as to make tools and parts for themselves. These practices led to duplication of effort which, though clearly more efficient at the very local level since they allowed plants to at least get something done, lacked efficiency at the macro level because they hindered both specialization and standardization. Standardization, or lack thereof, was a particularly important problem for manufacturers of such things as machine or automobile parts, but standardization across an industry was hard to achieve in an environment where each plant was manufacturing its own tools, often using makeshift machinery to do so. If plants were manufacturing parts using homemade tools, the parts might work independently, but they were not necessarily going to work well in combination with other parts made with a different set of homemade tools in a different factory. Better transportation could facilitate industrial coordination and improve both industrial output and the quality of products being produced. For this reason, transportation was a major preoccupation of both AWPM and CWPB staff.

Case Study: Transportation and Fuel

One of the most critical problems that the CWPB and AWPM sought to solve was that of transportation, a fundamental issue that Nelson had identified on his first mission as being key to fixing China's larger production problems. "If transportation cannot be improved," he said to Chiang Kaishek in September 1944, "China's economy is really in danger of collapsing."[49] Poor transportation networks, limited numbers of vehicles, and scarcity of fuel were a central problem for both industrial development and prosecution of military action. Massive limitations on transportation of supplies and equipment into China and of Chinese goods out of China caused one set of problems, but inadequate transportation of raw materials as well as finished goods within China caused another. As Nelson, Locke, and Jacobson learned from the FEA's Calvin Joyner in Chongqing in September 1944, the United States had flown in ninety trucks only a few months earlier of which thirty were already out of

FIG. 4.3. Transportation and road usage observed by Dykstra in northwest China. Courtesy of Harvard-Yenching Library of Harvard College Library, Harvard University.

repair, almost certainly because of the poor quality of the roads over which they traveled. "This gave an idea as to the rapid deterioration of equipment in China. He [Joyner] said that we need 40 times as much repair parts in China for truck maintenance as the army maintenance tables show are required."[50] LeRoy Whitney wrote to Donald Nelson just a couple of months later, stating the obvious, "The more I hear about the situation in China, the more I am convinced that transportation of raw materials and finished products is going to be one of the major problems."[51] Certainly, it was a problem of which both Chinese government officials and other observers spending the war years in western China were all well aware.

Every dimension of the transportation system was troubled. In the first place, although the ROC government had been actively constructing roads in the region since before the war, it was still the case by the latter stages of the war that relatively few roads in western China were well suited for automotive travel (see fig. 4.3). At the same time, there were very few railroads, and riverboats, though used extensively in some areas, were limited in their utility by the inadequacies of the road networks to which they were connected. Modern vehicles—cars, trucks, and buses—were also relatively scarce. The CWPB estimated that there were just over eleven thousand cars and ten thousand trucks (excluding military trucks) in the region in late 1944 and early 1945. Most reporters estimated that a minimum of 20 percent of all vehicles, and probably more like 30–40 percent, were off the road needing repair at any given time.[52] Finally, even in cases where roads and vehicles were not a problem, fuel was. There simply was not enough of it, and the fuel that was available was of widely varying quality and type and was typically very expensive.

Getting more trucks to China was one solution to the transportation problem, and both the FEA and the AWPM advocated for shipping American trucks to China not just for military use but also to move goods and resources within China. As Weng Wenhao, writing to Donald Nelson in March 1945, made clear, the production goals of the CWPB and AWPM simply could not be met without better transportation. "One of the major difficulties which we keenly experienced these few months in the promotion of war production is the lack of transportation facilities. For instance, in increasing the production of iron and steel, there is immediately the question of supply of coal, coke and iron ore which there is no sufficient means to transport. Also with power plants, facilities to ensure continuous and sufficient supply of coal remain a serious problem." What China needed, Weng went on to say, were trucks, and not just the castoffs that were used first for military purposes and would only be passed along to the CWPB later. "In order to render a substantial help to the different production units to facilitate increase of production, I wonder if you could help us by securing more trucks to be delivered directly to this Board for distribution to the various production units. If the transportation force of the various production units could be sufficiently strengthened, it will, I believe, greatly facilitate production."[53] This was not the first time trucks had been requested. Nelson had put in a request to President Roosevelt in October 1944 for an additional ten thousand trucks above the five thousand that were already at that time on order by the FEA, but as of April 30, 1945, few had been delivered, and the FEA had determined that the first five thousand to arrive in China would all be made available first to the U.S. military and only later to the Chinese government (an arrangement that was clearly not acceptable to Weng Wenhao).[54]

In spite of these pleas for more trucks, the number of trucks in China was still manifestly insufficient in the summer of 1945. In fact, the dearth of trucks was leading to problems in the distribution of materials coming into Kunming via plane over the hump. In early July, Lester Bosch, an electrical engineer with the AWPM, reported on a conversation with Dr. William F. Woo, director of the CWPB's southwest office, about truck needs in Kunming. According to Woo, there was such a backlog of items being brought in by plane over the hump that it would require one thousand trucks working for two months to distribute all of the material to its intended destinations and eliminate the backlog.[55] Meanwhile, the U.S. Army requisitioned more and more of the trucks that were being manufactured for China so that by July 1945, it was making the argument that the army should get exclusive use of nine thousand of the fifteen thousand trucks. Edwin Locke, by then head of the AWPM, vigorously defended the CWPB's need for trucks, but the point was essentially moot, as only 753 trucks had actually made it to Kunming by September 2, with another 8,183 still in various spots en route to China.[56]

Roads were, of course, another problem. The Nanjing government had undertaken to expand the number of miles of highway both within each province and connecting provinces, so that as of 1936 a geographer wrote that "with the exception of Xikang and Xizang, all of the remaining 28 provinces and districts can be reached by car," though he did not comment on the ease with which this could be achieved. Nonetheless, a new national highway network was being constructed that would facilitate government, economic integration, and national defense.[57] By the time the war broke out, Sichuan, in particular, had numerous internal roads, as did some of the other southwestern provinces, though not all of them had been built to standards that really supported the greatly increased volume of motorized vehicles that traveled over them during the war. Most of Sichuan's internal highways were built between 1926 and 1936.[58] In the northwest, however, there was a notable absence of roads that could be traversed by automobiles or trucks in the years leading up to the war. As Sir Eric Teichman, a British diplomat and China hand, reported of his driving trip from Beijing to Urumqi in 1935, there were three possible routes for him to follow: the old imperial road, which was essentially a cart track running through Shaanxi and Gansu, a caravan route through Outer Mongolia, and a camel road that ran through Inner Mongolia. As the first two routes were closed to him for reasons of political disturbance, he opted to follow the camel path, the most arduous part of which crossed the Gobi Desert.[59] Even modern roads remained problematic when traversing natural boundaries such as mountains and rivers. As Franklin Ho observed of road conditions he encountered on his northwest trip in the fall of 1943, "From Chengtu to Kwangyuan on the Szechwan-Kansu border, it was more difficult [than from Chongqing to Chengdu]. Roads had been built, but bridges had not yet been constructed across the rivers. We were forced to depend on the traditional ferries. . . . The worst part of the trip was the extremely mountainous and difficult countryside stretching from Kwangyuan to Tienshui in Kansu."[60]

The National Highways Bureau of the Ministry of Communications continued to build new roads during the war. Most of the highways connecting Sichuan with neighboring provinces to the north, south, and west, for example, were built between 1937 and 1944, but although a major road from Guizhou to Chongqing was completed around 1937, the other roads heading out of the province toward Gansu, Shaanxi, and Yunnan were all constructed during the 1941–1944 period.[61] Very early in the war, considerable effort was already being expended on road development in the northwest. J. Marvin Weller, a petroleum geologist who spent much of 1937 doing oil exploration in Gansu, Ningxia, and Qinghai, noted the fast pace of roadwork in Gansu. Returning to Lanzhou in December 1937 after having been off the beaten path in Qinghai and Ningxia for several months, he observed that during the time he had been gone, a bridge had been built and a great deal of other work had been done on the main road near Lanzhou, all with very low-tech traditional methods.[62] Not all parts of the Northwest Highway were built at such a fast pace, however. Shutang Lee confirmed that roadwork was still being done on this path in 1941 and 1942 but that topography was to blame for the slow pace at which transportation was improving. "In order to establish direct connection between Chengtu and Lanchow, a motor road is now in building, but owing to the topographical obstructions progress is slow. In many places mountain paths are so narrow and deteriorated that even mule traffic is impassable. Where highways are building, foot paths are also destroyed."[63] The construction of highways, thus, disrupted old transit paths that were traversed using traditional modes of transportation. Inevitably, old and new technologies tried to share the same road space, and it was not uncommon for road accidents to occur between motorized vehicles and pedestrians or cart animals.

By the fall of 1943, Joseph Needham, who experienced many of the same problems as Franklin Ho in the border area between Sichuan and Gansu, wrote that in Gansu, "immediately after one passes on to that section of it under the control of the Northwestern Highway Administration, a marked improvement is observed."[64] Whereas free China had about forty thousand kilometers of highways prior to the war, by the spring of 1944, the bureau was boasting that it had added ten thousand more kilometers of roads.[65] Some of the most important among these roads were the Xian-Lanzhou highway and the Chongqing-Urumqi (Dihua) highway, the latter of which was completed in the spring of 1944, as well as the Chongqing-Guiyang road, some treacherous parts of which were rebuilt in the early 1940s.[66] Along these new national highways, the bureau set up rest stations with hotels, shops, telegraph services, and perhaps most importantly, garages where vehicles could be repaired.[67] As Theodore Dykstra observed of these rest stations, where he stayed on his August 1943 potato expedition to the northwest, "Practically all the managers are university graduates and speak English." The manager of the highway department service station in Gwanyuan "was a university graduate in mining engineering and in 1926 went to America and spent 3 years in the Ford plants in Detroit."[68] These newly constructed service stations at which trucks and cars—new technologies for the region—stopped for rest or repair were important enough that they needed to be overseen by highly qualified technicians.

New trucks quickly became old trucks on China's poor roads, so truck parts and mechanics were just as important as trucks themselves. Given the condition of roads and the general scarcity of both automobiles and automobile parts, both the AWPM and the CWPB anticipated that maintenance of the trucks the CWPB would be getting from the United States would be of the utmost importance. One solution to the truck repair problem was to bring Chinese transportation experts to the United States for training. As part of the massive wave of Chinese engineers being trained in the United States in 1945, for example, a group of transportation engineers went in July for training in the bus shop of the

Capital Transit Company where, according to an article in the company's internal newsletter, they were particularly interested by the welding shop in which they learned about new tools that they might eventually employ to repair cars and trucks in China.[69] At the same time, the Chinese government and the FEA, with the help of Nelson, arranged in the spring of 1945 a contract with the Chrysler Corporation to set up a training program on truck repair and maintenance in China. The goal of the program was for Chrysler to set up fifteen repair shops in major cities and along the main highways at which a large number of personnel selected in the United States by Chrysler would undertake to both repair trucks and teach roughly three thousand Chinese trainees the fundamentals of car and truck repair.[70] By August 1, 1945, only thirty Chrysler trainers had departed for China, though the program was later taken over by the Chinese government and the contract with Chrysler renegotiated to better suit Chinese needs.[71]

Just as important as repairmen, however, were the parts they used. AWPM advisers and other Westerners in China commented frequently on the jury-rigged nature of many repairs and on the poor materials Chinese often had to use to make repairs, but different observers perceived these activities in very different ways. Observers such as Joseph Needham and Theodore Dykstra, both of whom spent a great deal of time on the road traveling around western China between 1942 and 1944, wrote extensively in diaries and letters home about the state of the roads, the frequent breakdowns, the impossibility of finding appropriate parts, and the need to come up with creative solutions.[72] In their telling, the solutions to these problems required ingenuity and resourcefulness, two things that virtually any "technical man" in western China had to have in order to achieve results in primitive conditions. Both Needham and Dykstra were thoroughly impressed by the ingenuity and resourcefulness that they encountered there.

AWPM technical advisers, however, looked at the problem from a very different perspective. Their concern was with establishing standards, increasing production of quality goods, including auto parts, and improving distribution of those goods. For them, solutions should not have to depend on resourceful jury-rigging. Rather, auto parts factories should manufacture quality parts made to a set of national standards that would be distributed to the appropriate places and available when needed. This would not only make auto repair more predictable, effective, and possible, but it would also have positive effects elsewhere in the industrial system as factories manufacturing other things would no longer be required to diversify their production by manufacturing their own substandard parts to keep their trucks running. Beyond these immediate benefits, such improvements would also bring China more thoroughly into an increasingly globalized economy. Chrysler was already making inroads in China by supplying not only trucks but also technicians and training, all of which would position the company to continue to engage with China after the war. Other efforts to modernize China's infrastructure and industry using American standards, systems of thinking, and products would have the same effect.

There is clearly some tension here between the ideal goals of the CWPB and AWPM advisers on the one hand and the real conditions in China on the other hand, and this tension can be seen not just in the contrast between the observations of different kinds of advisers noted above but also in the writings of AWPM advisers themselves. Those advisers spent much of their time on the road doing industrial site visits and as a result they were fully aware of the hard usage that trucks and cars received in those conditions, of the relative isolation of many factories and industrial sites from each other, and of the reasons why factories did choose to diversify and to manufacture their own parts. At the same time, they were in China on a mission with clear goals. It was their job to make recommendations that would facilitate the accomplishment of those goals, and one of those recommendations was that plants should focus on the things they were supposed to make.

The CWPB shared these concerns. Auto parts were of sufficient importance to the CWPB that it had an Auto Parts Advisory Committee. The work the CWPB had done toward standardizing manufacture of auto parts was described on the agenda of the committee's second meeting in March 1945. "Except few works, usually there is no specification for manufacturing and also no inspection code. A few war emergency standards of auto-parts is [sic] being worked out jointly by the [Chinese] WPB and the Industrial Standardization Committee. Besides, WPB has worked out tentative specifications and inspection methods for all articles to be ordered this time. Hoping the parts manufactured hereafter will be up to standard always." The CWPB, however, remained concerned that transportation organizations were making "lots of parts themselves in their repairing shops. It has serious effect on the quality of the products and the proper maintenance work of trucks. This needs immediate attention."[73]

This is not to say that good auto parts were not being manufactured in China. Ralph Strang, a machine tools expert with the AWPM, particularly commended the China Automotive Works for having "produced one of the most accurate small lathes that have been made in China. They have also produced automotive parts of very high quality," and the Central Auto Parts plant, of which he wrote that "the truck and automobile springs produced by this firm kept hundreds of trucks on the road. Without the springs the trucks would have been entirely out of commission which would have resulted in the loss of transportation running into the thousands of tons. The piston rings produced by them was another item that kept many cars and trucks in operating conditions."[74] Such parts were keeping cars and trucks operational, and it was clearly in the interest of both the AWPM and the CWPB to see that China's auto parts manufacturers continue to develop the range of parts that they could produce and distribute across western China.

The third major transportation challenge, certainly as important as poor roads and insufficient and inadequately maintained vehicles, was fuel. Prior to the war, petroleum had been entirely imported. As one geologist (clearly located in coastal China) wrote in 1935, "As a result of the wide industrial development as well as extensive construction of the public highways and the consequent demand of motor-car gasoline, the petroleum consumption in China is becoming greater and greater every year. According to the recent report of the Customs office, the import of gasoline, kerosene, fuel oil and lubricating oil has reached in 1933 a total of about eight million barrels, valued at 140 million Shanghai dollars." He went on to observe that "the present situation of depending entirely on the foreign oil supply means not only a great loss to our national revenue, but also a great danger to the country, because petroleum is such an important war mineral." This foresighted author went on to urge both the development of a Chinese oil field and investigation into alternative fuels that could replace petroleum.[75]

Only a couple of years later, China was indeed cut off from petroleum imports and compelled to fast-track the development of the Yumen oil field in Gansu and also to start manufacturing several alternative fuels, the most successful of which was power alcohol, to meet demand while wells were drilled and refineries constructed. Moreover, as the Yumen oil field was located on the wrong side of the slow-going and treacherous road that ran between Lanzhou and Chongqing and had neither sufficient trucks nor appropriate equipment to transport very much oil to Sichuan or farther south, even once the field was producing at a relatively good rate in 1943, its impact was most heavily felt in Gansu and the northwestern provinces. Sichuan and the southwestern provinces were compelled to rely almost entirely on alternative fuels. Most alternative fuel was alcohol produced from gaoliang or sugar, although some was produced by cracking vegetable and tung oil, and in addition, some vehicles were adapted to run on charcoal.

Owing to the scarcity of fuel as well as the fact that the most abundant fuel in Sichuan and the southwest was manufactured from foodstuffs, the Chinese government controlled fuel production and distribution through an LFCC that it established in 1938. The LFCC allocated production to the various distilleries, set prices for finished products, and determined who could purchase how much fuel.[76] As the chief entity in charge of fuel production and distribution, the LFCC had considerable authority, and its decisions regarding allocation of contracts and fuel had a significant impact on the degree to which plants were able to operate at capacity (or, indeed, operate at all) and on whether applicants for fuel (all of which were either government or private enterprises rather than individual consumers) were able to conduct their business or transport their goods.

Sichuan's first industrial alcohol plant was set up in 1938, soon after the outbreak of war and as the need for alcohol as a fuel was only just beginning to become apparent. Between 1939 and 1942, the number of alcohol plants increased rapidly so that by 1943, there were sixty-one plants producing industrial alcohol in the province. Until 1941, both sugar and molasses, the latter of which had often been thrown away prior to the war, were cheaply available, and as a result, most of the early plants were set up in Sichuan's sugar zone along the Tuo River between Chongqing and Chengdu. In 1941 and 1942, however, crop prices began to increase, acreage devoted to sugar production declined, and the prices of both sugar and molasses went up, leading to a sharp decline in the number of operational plants. Along roads in other parts of Sichuan, numerous alcohol plants were constructed that were originally intended to distill alcohol from grains or potatoes, though as it became clear that they could not get these materials in the quantities they required, most of these plants eventually turned themselves into secondary distilleries of native wines. Just as Sichuan's alcohol boom began to settle down in the early 1940s, Yunnan and Guizhou began to build new plants, with the greatest number of plants being constructed in 1942 and 1943 in both provinces so that by 1944, Yunnan had thirty-nine plants and Guizhou had seventy.[77] In spite of the challenges the developing industry faced, between 1940 and 1944, private industry was typically manufacturing more than twice as much as state-owned industry and alcohol production steadily increased from about 5.5 million gallons in 1940 to nearly 8.7 million gallons in 1944.[78]

Fuel production, though not really sufficient to meet the needs of the Chinese government, military, and industry, was deemed to be sufficient by the LFCC because that entity was most concerned with protecting the balance between food availability and fuel production. From the Chinese perspective, alcohol was not the ideal fuel precisely because it was made of things that could be eaten. Although the vast majority of fuel alcohol was produced through redistillation of "white wines" that were most typically made of sorghum (gaoliang), some was made directly from sugar and molasses. Sorghum, though most commonly used for animal feed, was a secondary grain that people consumed in times of need. Sugar and molasses were also foodstuffs. Acutely aware of this tension between fuel for machinery and fuel for people, Chinese leaders were inclined to emphasize the development of their petroleum industry, which did not place any burden on food production.[79] Requests for U.S. assistance prior to the establishment of the CWPB repeatedly mentioned the need for equipment to improve both extraction and processing of Yumen oil and reminded American advisers that potential alcohol production was limited by availability of raw materials.[80]

From the AWPM perspective, however, there was simply not enough fuel in China, and alcohol was the only immediately viable fuel that could be used to propel the trucks that they were starting to ship into China in 1944 and 1945 to assist with the anticipated ground offensive against Japanese forces. Alcohol was also the best fuel produced in China for airplane use. Armed with an estimated demand by the U.S. Army of thirty million gallons of fuel alcohol, the AWPM prioritized alcohol production, although they also offered advice and some material assistance to other fuel development efforts. To that end, one of the first advisers sent to China by the Nelson mission was Eugene M. Stallings, an industrial alcohol consultant who had been chief of operations with Joseph E. Seagram & Sons Company since 1936 and who had worked for the U.S. War Production Board since 1942. Aware of government concerns that alcohol production not lead to reduction in available foodstuffs but apparently not fully understanding, at least at first, that sorghum was not solely used for animal feed,[81] Stallings encouraged alcohol manufacture, particularly from native wine made mostly of sorghum. Throughout the spring of 1945, however, the food issue remained a concern, and Harold Roland, who succeeded Stallings as alcohol expert in the spring of 1945, maintained correspondence with various representatives of the FEA in Kunming about the impact of alcohol production on food availability. Although it is apparent from this correspondence that not all members of the FEA agreed with him, Roland ultimately concluded that Chinese government statistics showed that producing thirty million gallons of alcohol a year would use "less than 3% of all grains grown" and continued to argue in favor of pushing for increases in alcohol production.[82]

Convinced that the food issue was not significant, Stallings and Roland both argued that the greatest limitations on production were economic and structural. They saw the LFCC as an obstructionist entity and were dismayed by the CWPB's failure to compel it to make any real changes in its methods. Throughout the winter of 1944–1945, Stallings made repeated complaints to his superiors, and through them to both Weng Wenhao and T. V. Soong, about the LFCC's intransigence and unwillingness to expand alcohol production.[83] To remedy the situation, the AWPM urged the CWPB to create a more streamlined institutional structure that gave them full authority over all dimensions of the fuel production process.[84] To that end, the LFCC was brought under the authority of the CWPB, but in spite of this move, Stallings was particularly distraught that even with Weng Wenhao present at a meeting to discuss alcohol production, nothing had really changed. In fact, Stallings felt that "the meeting tacitly gave approval for LFCC to continue to function as formerly with only casual and desultory control by CWPB."[85] It seems plausible to conclude that one reason that the LFCC continued to operate as it had was that although Weng expressed his support for the alcohol plan to AWPM advisers in order to appease them, in fact, he continued to have concerns about food availability and was not really eager to eliminate the controls that the LFCC had in place. A different possibility, however, is that Weng was simply not able to bring the LFCC fully under CWPB control. Either way, this story illustrates the power that AWPM advisers had over the CWPB and the influence that their views had on its organizational structure. But it also demonstrates that the Chinese actors whom the AWPM sought to influence did not always fully comply. In this case, although the institutional structure was changed, the LFCC continued to operate as it had before.

In addition to the structural problem, both Stallings and Roland noted a number of economic obstacles to fuel production, many of which they believed could be overcome by actions on the part of the CWPB. In the first place, they found that the formula for fixing fuel prices disadvantaged producers by not allowing for increasing commodity prices. Fuel prices for a given month were determined by a formula that took into account the previous month's cost of coal (to fuel the stills) and the cost of raw materials (native wines, sugar, molasses, or grains). This formula failed to take rapidly rising commodity prices into account. Thus, distillers frequently lost money or lacked capital to purchase enough raw materials to operate at full capacity. This problem was further exacerbated by high interest rates on loans.[86] A further problem was that the LFCC set different prices for different regions, leading to unnecessary (but economically lucrative) transportation of fuel from cheaper to costlier producing areas, thereby using valuable fuel to move fuel around rather than conserving it for use by LFCC designees. In addition, different regions had different regulations regarding the use of foodstuffs to produce fuel (Yunnan, for example, did not permit the production of alcohol from food items but did not count sugar as a food item), and the LFCC had done nothing to remedy these disparities. At the same time, the permit system operated by the LFCC led some large users to be granted more alcohol than they needed, which had the consequence of both leaving small users unable to get alcohol from the LFCC, and allowing the large users to sell off their extra on the black market at higher prices.[87] Also problematic was the LFCC's contract system. Stallings argued that the contract system "will not accomplish the desire to pull the industry together all working for one aim—to win the war. It still leaves the idea that the main object is to make money or to purchase alcohol at as low a price as possible. The favored distillers will make money. The less favored will still continue to struggle along in a disgruntled and unpatriotic frame of mind." To solve all of these problems, Stallings proposed development of a new system of universal pricing and the extension of financial assistance by the CWPB to producers that would provide them with the capital they needed to operate at full capacity. The LFCC (or better yet, the CWPB) should purchase alcohol at cost plus profit but should sell at a single price (the average price) throughout China.[88]

Transportation, though by no means the only area of focus for the AWPM, is a good case for understanding overall AWPM goals and the disjuncture between those goals and the realities of wartime inland China. It also shows us the deep interconnectedness of the work being undertaken by the various working groups of the CWPB. Widely dispersed factories had limited access to raw materials, they were heavily dependent on trucks to supply them with such materials, trucks were dependent on factories to produce both parts and fuel, and the roads over which the trucks traveled were generally inadequate. Each of these pieces alone would have been a massive project to fix in wartime inland China, but the CWPB and the AWPM were looking to find ways to overcome all of them at the same time.

In general, AWPM technical advisers took an on-the-spot approach to helping factories to fix technical problems that would help them to increase production. However, the case of transportation underscores the importance of macrolevel planning as well. No matter how well the machinery in any given plant could be made to function, the plant would not be able to operate at capacity without raw materials, and its output would be meaningless if it could not be gotten to those who needed it. To make all of this work required a coordinated effort.

The case of transportation also shows the heavy importance that the AWPM placed on standardization as a strategy for achieving industrial efficiency and increasing output. With standardization of production and the construction of an effective distribution network for the standardized parts, they expected that trucks would last longer and be better able to move raw materials and finished goods around western China. This, in turn, would both serve American military interests and help to stabilize production, which AWPM advisers believed would also help stabilize the economy.

Finally, the case of transportation and fuel demonstrates the heavy degree to which American advice was guided by American interests. Automotive consultants were guided at least in part by the desire to establish industrial relationships and consumer expectations that might serve U.S. business interests in China after the war. With regard to fuel, the fact that AWPM advisers focused first and foremost on alcohol rather than other biofuels or petroleum is clear evidence of the American emphasis on short-term solutions to meet immediate wartime needs. Development of the Yumen oil field, though a major point of focus for the Nationalist government, was essentially dismissed by the AWPM as it would take far too long to get anything useful out of it that would support the war effort. Stallings's continued insistence that thirty million gallons of industrial alcohol was an achievable and appropriate goal for 1945 also attests to the fact that he was guided by American military interests. That production goal took *only* the U.S. military's need projections into account and ignored the fact that much of the alcohol produced in prior years had been distributed not to the military but to Chinese government organs and enterprises as well. In his final report, Stallings proposed that to serve the prosecution of the war, "all non-essential use of motor transportation should be banned. The transportation of people in trucks suitable for cargo should be minimized. It should be unlawful to possess, sell or purchase liquid fuels without authority from the [Chinese] War Production Board."[89] In other words, the prosecution of the war was his most important goal as he doled out advice regarding alcohol production.

The case of transportation and fuel also shows us where CWPB interests could diverge from those of the AWPM. Although the CWPB was animated by many of the same concerns as the AWPM, its leaders and members had much more context for understanding the challenges of the tasks they sought to accomplish. Their end goals were also not precisely the same. Some AWPM advisers were, as noted above, motivated by postwar economic interests, but the ethos of the institution was the one articulated by Stallings as he pushed for production to meet military needs. CWPB leaders and members were, of course, thinking about the short term, but they were also thinking about the larger context. Production needed to serve nonmilitary as well as military needs and industrial development needed to be done in a way that would endure beyond the war.

As the discussions over alcohol production make clear, CWPB members were not always willing to follow through on AWPM advice.

Conclusion

The very creation of the CWPB demonstrates an effort by American advisers to impose an American industrial/bureaucratic ethos onto China. As Mabel Gragg observed in her "History of the American War Production Mission in China," written immediately after the termination of the mission, "What the Mission in effect undertook to do was to modify the traditional Chinese attitudes and ways of doing things in the direction of American attitudes and ways."[90] In promoting American industrial norms, standards, and products in China, the AWPM also worked to pull China into an American industrial orbit. Greater efficiency and increased production were the mantras by which AWPM advisers operated, but although the CWPB was created and given considerable authority, it either could not or would not exert that authority in every way that AWPM advisers wanted. It is clear that AWPM advisers did seek out as much information as they could get on real conditions, but even in spite of that, their expectations of what was possible did not entirely match the reality of the political, economic, and social situation.

The CWPB and the AWPM were a logical outgrowth of the kinds of technically oriented relationships that were initiated with the State Department's cultural relations program. Guided much more explicitly by American military interests and Chinese economic development interests than the cultural relations program had been, the two new entities created a new model for Sino-American scientific and technical collaboration. Unlike most of the wartime strategies to promote scientific and technical development that Chinese bureaucrats had developed up to that point, the AWPM and, by extension, the CWPB, were focused almost entirely on the present. They were created to address immediate problems using available resources and to improve the situation by marshaling, to the extent possible, additional resources from abroad. The American technical experts who constituted the staff of the AWPM conducted careful and apparently thorough surveys of industrial conditions, provided on-the-spot technical assistance and advice, and developed plans aimed primarily at improving industrial efficiency and productivity. CWPB staff worked to streamline bureaucratic processes, identify solutions to technical problems, and make investments that would increase production of strategic commodities. In general, they did succeed in increasing output of key products.

However, it is also clear that Weng Wenhao and the CWPB staff were acutely aware of the temporary nature of conditions in wartime China and still thinking about the longer term. Even as they agreed to the many suggestions made by Nelson and the AWPM staff, they were always thinking about what more China would need to actually modernize Chinese industry and not just to win the war against Japan. Some of the work of the CWPB, such as efforts focused on setting standards, had the potential to have long- as well as short-term impact, but ultimately, what Weng and the CWPB most wanted out of the technical relationship with the United States was industrial machinery and equipment and engineering advice that would help them position Chinese industry to take off quickly after the war. As Gragg noted, "The Mission should be credited not only with raising output approximately 25% over a six-month period, but also with having checked a decline which was rapidly reducing Chinese industry to impotence."[91] Even accepting the possibility that Gragg's assessment might have been a bit exaggerated, there is little doubt that the AWPM and the CWPB worked together to turn around what had been a decline in the output of strategic raw materials and industrial manufactures in 1944 and to increase production in the spring and summer of 1945 in a number of key areas.

Perhaps the most important role of the CWPB was that it provided yet another umbrella under which Chinese technocrats could build and cement relationships with American technicians, businesses, and bureaucrats that could be employed to China's advantage in the postwar era and through which AWPM advisers could do the same. As soon as the war ended, former AWPM staffers were already jockeying to profit from China. LeRoy Whitney, who had resigned from the American War Production Board just days before the end of the war to open an export-import company is one good example. In a letter to Hollis H. Arnold of the FEA, Whitney wrote, "I am very mindful of the importance of export business to the United States, and I also believe that if we are going to sell our goods to foreign countries, we have got to help them sell to us and to others. The reconstruction and stabilization of the world looks to me to be a much bigger job than winning the war, and our little group here in Intercontinental Distributors, Inc., hope to help the situation as well as to make a little money out of it." To accomplish this goal, Whitney wanted to capitalize on the relationships that he and others had built in China while working for organizations such as the AWPM and the FEA. "We are particularly interested," he wrote, "in setting up an organization to which the Chinese government and Chinese industry can appeal for advice regarding the selection of consulting engineers, etc., as well as an organization which will represent an outstanding group of American manufacturers in China."[92] Whitney's letter serves as stark evidence that American motives for wartime engagement with China were just as complex and multifaceted as Chinese motives for engaging with the United States. Both sides wanted to win the war, for sure, but every actor on both sides of the relationship was also motivated by other concerns.

The CWPB was relatively short lived. The war ended nine and a half months after it was created, and the CWPB remained operational only through the end of 1945. Edwin Locke, who took over leadership of the AWPM in the summer of 1945 and who was employed on special projects in the Truman White House, maintained an ardent interest in China's economic development well after the CWPB was shut down and kept in close contact with a number of Chinese bureaucrats as well as American businesses that were engaged in work with China. The nature of the technical advising relationship that had developed during the war changed in various respects in the months and years immediately following the war, but those changes were nonetheless built on foundation of the AWPM along with the wartime training programs managed by the China Institute. Chapter 5 will examine Sino-American technical cooperation in the immediate postwar era.

CHAPTER 5

Training, Planning, and Reconstructing the Transnational Relationship in the Postwar Era

The war had ruptured many existing patterns of production, trade, resource development, international engagement, and government and compelled Chinese actors and their Western counterparts to develop new approaches to all of these things, all while dreaming that the moment the war ended, certain old patterns would resume, and others would be replaced by new ways of doing things that had been meticulously planned during the war. In some respects, however, at the end of the war, things just kept going as they had during the latter stages of the war. Certain patterns of international engagement that had developed during the war, especially with regard to international training programs, continued. In other respects, such as U.S. technical advice, new patterns emerged that better suited the changing nature of the Sino-American relationship and the needs of the Nationalist government. In still other ways, the end of the war was almost as massive a rupture as the beginning of the war had been. This chapter examines the ways in which Chinese and American approaches to training, technical advising, and planning evolved in the immediate postwar years and considers the impact on these things of changing circumstances and motivations of the actors.

Shifting Away from Inland China

Development of inland industry and agriculture to serve the war effort and to support the increased inland population had been a major priority for both the Chinese government and its American advisers during the war. As we have seen, these development efforts were undertaken within a set of extraordinary constraints including absence of imported goods, lack of access to export markets, limited access to domestic or imported raw materials, poor transportation infrastructure, absence of competition, and a compelling need for government intervention to manage both industrial and agricultural output. The end of the war radically transformed the context for both Chinese scientific and technical planning and development and Sino-American scientific and technical collaboration. With the end of the war, both the Nationalist government and their foreign advisers swiftly turned their attention to newly liberated territories with the expectation that the industrial and agricultural capacity of these territories, which had been much stronger than that of inland China prior to the war, could be quickly rehabilitated to serve the needs of the newly reunited Chinese nation. This refocusing of attention away from the inland provinces and toward coastal China, Manchuria, and Taiwan posed an interesting dilemma regarding China's inland provinces. Should China's planners and advisers turn their backs on all that had been built in the inland, or should they try to improve on the foundation they had built, continue to develop the region, and support the industry that remained after the war?

In spite of the fact that planners, as we saw in chapter 1, universally planned for the integration of China's western provinces into a national economy, in most areas neither industry nor infrastructure had yet been adequately developed to make that possible. Even within the local economies of inland regions, factories that had been constructed during the war did not appear to many observers to be sustainable in the postwar period given the costs of doing business in the region. As W. A. Haven of the Arthur G. McKee Company noted in a May 1946 report on iron and steel plants in the Chongqing area, "Any planning for the industry in the Chungking District must begin with the development of cheaper methods for transporting raw materials."[1] The cost of transporting the raw materials within Sichuan that were needed for the manufacture of steel in Chongqing was so high, in Haven's view, that just this one issue made Sichuan steel manufacturing, even for the local market, seem to be impractical, and it certainly priced products manufactured in Chongqing out of the national market. The proliferation of iron and steel plants, many of them small, in the Chongqing region had made good sense in the context of the war, and even the additional manufacturing costs had been justifiable under wartime circumstances. In a postwar world, given the infrastructural challenges within the region, their future prospects seemed dim. "Obviously," he observed, these plants "cannot operate during peace times under such conditions."[2]

The economy of inland China during the war operated under a set of conditions (no or few imports, limited markets, no competition) that acted as both challenges and opportunities. New industries grew that served local and military needs, government organs undertook to provide research and development support and to develop standards, and the Nationalist government strove to increase its control over inland industry. For as much as the war had provided the opportunity for these sorts of changes, however, it also forced industry to develop in ways that it would certainly not have developed under other circumstances. As the war ended, the wartime conditions that had so challenged production, though somewhat alleviated, did not change dramatically. Despite the reopening of land and water routes between inland and coastal China, transportation, especially over land, continued to be a huge impediment to both the importing of resources to the hinterland and the exporting of manufactured goods and raw materials from those same areas. Given that the government was no longer compelled to focus all of its economic development energies on this territory, these continued constraints and emerging challenges all combined to make continued focus on the development of inland industries after the war less attractive. Many of those industries were no longer as critical to the region as they had been during the war, and the costs of both manufacture and transportation of goods produced inland meant that their products were not necessarily competitive in markets elsewhere in China.

Not all inland projects were abandoned. Some of the work that had begun in inland China during the war continued in the postwar period. As Raymond Moyer, a member of the 1946 U.S. agricultural mission to

China, noted in a report on a trip he and other mission members made to National Northwestern Agricultural College in Wugong, Shaanxi, "This visit gave us our first opportunity to observe a part of unoccupied China, and we were struck at once with the contrast between it and what we had seen in the parts recently liberated. In formerly occupied territory work was disrupted and still in the early stages of being reundertaken. Here we saw a going program, and we gained a better conception of what has been done for agriculture under the Central Government."[3] Similarly, the National Resources Commission (NRC) continued to work on further developing certain inland industries that were of strategic importance, such as iron and steel factories in Chongqing and the oil industry in Gansu. In general, however, the conditions that had been created by the war began to change in ways that were not necessarily beneficial to the continued economic development of the hinterland. In particular, the large population of educated coastal elites with academic and/or technical knowledge who had participated in and, in many cases, guided the economic development of the region during the war gradually worked their way back east to their home cities and provinces, or abroad. The distance between coastal and inland provinces was made even more apparent by the scarcity of good transportation networks that could move raw materials, finished goods, and people between inland and coastal regions. In many respects and in spite of the integrated national plans that Nationalist planners had developed during the war, the Nationalist government and the many coastal technical experts who had followed it refocused their attention on coastal China and, in doing so, sometimes abandoned successful programs that had developed in the straightened conditions of inland China during the war.

The Nationalist government's optimism that Japanese-controlled industries could be quickly rehabilitated was in many cases ill founded. As it took over such industries, the government found that many prewar factories had been looted for their machinery by the Japanese and were consequently incapable of production. Even factories in north China that had operated under Japanese control could not easily be brought back to life because they were hampered by general economic or political instability, shortages of fuel and power, and/or a general lack of skilled managers and technicians. At the same time, physical destruction of housing and inattention to infrastructure such as sanitation, health care, and education in some areas under Japanese occupation all meant that refugees often found themselves returning to very different conditions than those they had left.[4] Government planners and technocrats, along with their foreign advisers, thus found that recovery and rehabilitation of Japanese-occupied areas might not flow so smoothly after all. Two regions that attracted the most attention were Manchuria and Taiwan, both of which had been developed under long-term Japanese occupation and both of which were better positioned with respect to availability of natural resources and access to transportation than inland China. As it turned out, recovery of Manchurian enterprises was not nearly as straightforward as the takeover of enterprises in Taiwan, but those complications were not foreseen as the war ended, and planning for various dimensions of that takeover continued well into 1949 even though prospects for a Nationalist recovery of Manchuria surely must have looked incredibly bleak by that point.

After the war, the Nationalist government continued to seek technical assistance from the United States to support these rehabilitation projects, but with the end of the war, the mechanisms through which such support could be sought and provided changed. Moreover, as time passed, the U.S. government grew increasingly concerned that because of political and economic instability in China, it could not effect the same sorts of rehabilitative change there that it was achieving in Europe through the Marshall Plan. This chapter examines technical assistance programs that spanned the late-war and postwar periods as well as new approaches to getting technical assistance and help with planning that the NRC developed after the war. It shows how, with a declining level of U.S. government involvement in providing technical assistance, the NRC took an increasingly active role in soliciting, guiding, and shaping the technical assistance it got. To that end, the NRC worked directly with private American firms to help plan for China's redevelopment and build a case for continued U.S. government assistance in this process. This chapter also explores the motivations of those firms for participating in such activities.

Technical Training

The anticipated takeover of Japanese-controlled factories would require an enormous number of Chinese managers and technicians who understood not only the technical dimensions of each industry but also how to rehabilitate damaged plants, and what organizational systems and personnel management strategies were needed to keep a plant operating at maximum efficiency. The Nationalist government and its American advisers developed a two-pronged strategy to produce the necessary personnel: sending large numbers of trainees to the United States and bringing U.S. managers to China. The former was initiated, as we have seen in chapter 3, months before the end of the war in hopes of building a workforce that could serve wartime needs in some fields and be ready to work as quickly as possible after the war in other fields. The latter was discussed and, in some cases, arranged for starting in the last months of the war and in the years immediately after but was not generally well implemented because the Nationalist government was not able to make the inroads it had hoped to make in Manchuria and also because it was unable to secure from the United States the funding it needed to support the projects. Both strategies aimed at expanding the number of skilled technicians and managers who could oversee and implement an extensive program of industrial and agricultural revitalization.

As we have already seen, China's overseas training program was ramping up even before the war ended, with its first very large cohort of 625 trainees arriving in the United States in late June 1945 and a second large group arriving in late September. Trainees were selected through competitive examinations in which they had to demonstrate, among other things, their command of English. They were all college graduates with at least two years of work experience. Many were drawn from NRC factories and institutions, but some were college graduates in technical fields who did not yet work for the NRC.[5] After being selected and preparing to travel, trainees left China over the hump to India and then traveled by boat for a month to Norfolk, Virginia or New York, from which points they went directly to Washington, DC. The program, which was initially sponsored by the U.S. Foreign Economic Administration (FEA), was taken over by the Nationalist government's Ministry of Economic Affairs (MEA) following the end of the war. The number of trainees continued to increase not only in the final months of the war but over the course of the couple of years following it. As peace broke out, travel became easier and less perilous, and the need for trainees in a diverse array of specialized technical fields and also in business and industrial management became even greater.

To a large extent, the patterns and networks that had been established through the earlier training programs described in chapter 3 (those that took place in 1943 and 1944) served as the model for this large-scale training program in 1945 and 1946, though the institutions organizing the trainings and placements had changed. Whereas the Universal Trading Corporation (UTC) and China Institute had played key roles in managing many of the earlier trainings, the FEA/MEA training was a collaboration between the FEA, International Training Administration (ITA), China Supply Commission in Washington, DC, and the NRC, which had offices in both Washington, DC and New York. Most of the trainees were associated in one way or another with the NRC or its various enterprises, and the NRC in New York (NRC-NY) maintained correspondence with many of them. The ITA managed and regularized certain dimensions of the program, so most trainees undertook an orientation at Georgetown University for the first month of their program and remained there while the ITA, often working in conjunction with staff at the China Supply Commission, sorted out on-the-job training sites.

U.S. PARTNERS

Both the orientation and the job placements drew on existing connections that had developed during the war between the Chinese government and U.S. companies and individual experts. At the outset, the China Advisory Branch Committee of the FEA helped to expedite the program and find placements for trainees. Among the committee members were seventeen representatives from industry, including the auto, oil, communications, steel, and banking industries. J. Franklin Ray, chief of the China Division of the FEA, asked committee members themselves to accept trainees in their factories.[6] The ITA, the board of which consisted of primarily industrialists, also drew on its extensive connections to assist with placement. Both groups already had experience interacting with the Chinese government and its representatives. Deeper connections between China and the United States were drawn on as well. Paul B. Eaton, for example, who had spent time in China as part of the cultural relations program (see chapter 2) and who had in 1943 facilitated the training of a group of Chinese trainees at Bell Telephone Laboratories (see chapter 3), oversaw the orientation for industrial management trainees in Washington, DC.[7] And placements were often made with government agencies and companies that had some existing relationship with the Chinese government, such as the Tennessee Valley Authority (TVA) and International Harvester, where other Chinese trainees had been placed earlier in the war.

An important motivation for U.S. participation in the training program was a belief that one of the United States' most important exports was know-how, and U.S. companies were eager to share what they viewed as a fundamentally American approach to getting things done and, in doing so, to expand the market for American goods. Not unlike Nelson and some of the AWPM advisers, American industrial consultants were often bent on molding China's industrial culture to fit an American image.[8] Glen Ruby, of Hoover, Curtice and Ruby, James Pierce, of Pierce Management, and Edwin Locke, for example, clearly believed that American know-how, including American industrial values and modes of doing, could be applied in China. To a large extent, that know-how would be transferred by establishing strong training programs and putting people through them. But know-how, as they used the term, was also about developing plans, figuring out the right tools, and putting both to use.[9] Perhaps most importantly, know-how was also a justification for the extension of American business interests into China.

Ruby, Pierce, and Locke were by no means the only Americans thinking about the merits of exporting know-how to China and elsewhere around the world. At precisely this time, and no doubt influencing the way in which American consultants framed and described the significance of their activities in China, a movement to export American know-how was growing in the United States under the auspices of the ITA. As the war drew to a close in both Europe and the Pacific, American leaders were looking to set up mechanisms that would foster peace and facilitate the extension of American political and economic influence around the world.[10] Eliot Hanson, head of the ITA, was a leading advocate for the role that the exportation of American know-how could play in the quest to achieve these aims. As Hanson told the American Society of Mechanical Engineers in a November 1945 speech, the United States, believing that industrialization and a rise in the standard of living would help bring long-term peace to the world, had, during the interwar period, eagerly exported capital and goods without providing the framework within which those things could be used or the know-how to use them appropriately. This approach, according to Hanson, resulted in unproductive operations and defaulted bonds and did not, in the end, prevent the coming of a second world war. As Hanson observed, "There cannot be worldwide industrialization with capital and goods alone. It takes plenty of know-how too! And today the world's richest source of this precious commodity is right here in the U.S.A. . . . Of such know-how we have an abundance, circulated without restriction in the domestic market. If exported with freedom, it will help to promote American interests abroad."[11] Those interests were both political and economic. Hanson, when describing ITA's training role, made sure to emphasize the potential impact that such training in American industrial methods and standards might have on trade. "In the coming surge of world-wide industrialization," he observed, "we will do well to insure that American standards and practices predominate in so far as possible. Thus will be sown the seeds which will mature into a ready market for American products."[12] Hanson drew a very clear connection between the export of know-how and the creation of markets for American-made goods. The United States was well positioned, in the wake of World War II, to capitalize on the weaknesses of other economies and to take a leading role in the development of economies around the world. Hanson and others may have cloaked these economic designs in a language of freedom and friendship, but ultimately, the goal was to help American private enterprise to make inroads into new markets.

This idea—that export of know-how could both foster peace and promote trade—clearly resonated with American businesses. As one article in the Coca-Cola Company's publication, *Red Barrel*, asserted, "Quite likely it is safe to say that the training of foreign nationals in the United States may prove to be one of the most effective instruments for promoting international trade with the United States and developing those operations of a peace-loving world which we seek so earnestly."[13] A Merrill Lynch publication put it even more directly, observing of training programs for Chinese technicians that were being run through ITA, "All this has a very practical side—if Chinese technicians learn the modern ropes on US railroads and US equipment, China is far more likely to buy American-made goods when the time comes. But more important is the simple fact that common-sense programs like this are a marvelous contributor to peace on earth."[14] U.S. businesses imagined themselves guiding the world toward a lasting peace through international trade. This point was also not lost on Chinese who were attempting to arrange placements for trainees. As one Major Chen wrote to S. C. Wang 王守兢 of the NRC-NY in September 1945, "It may not be a poor policy for any one of the companies to train us since a program like this will naturally carry with it a great deal of good will of the firm right to China."[15] In other words, Chinese trainees would make great ambassadors for the U.S. businesses in which they trained.

As the NRC developed relationships with additional U.S. firms through other consulting mechanisms, those firms also assisted in the creation of programs for the growing number of Chinese trainees heading to the United States under this program. Pierce Management, a Scranton, Pennsylvania firm that specialized in coal mining, serves as a good example. The relationship between Pierce and the Chinese government dates from mid-June 1945, when Alex Taub, a former FEA engineer who was now with the China American Council of Commerce and Industry and a consultant with the China Supply Commission (CSC) in Washington, DC, recommended that James Pierce, president of Pierce Management, connect with the CSC with an eye to setting "forth our views on the rehabilitation and development of China's coal industry with the thought that these views might prove helpful before the departure of Mr. T. V. Soong for China."[16] The NRC subsequently sought to engage Pierce's assistance in advising on coal development in China as well as the purchase of specific mining equipment to be used in China after the war. Pierce continued to provide consulting services on coal mining for the Nationalist government through 1948. In addition to serving as a consultant, Pierce helped the Nationalist government in other ways. In September 1945, James Pierce had become sufficiently connected that he conducted a training session for Chinese coal mining engineers at Georgetown. By the following July, Pierce Management had overseen the training of twenty-two Chinese technicians, nineteen of whom were studying mining and three of whom were studying industrial management. For the mining group in particular, Pierce served as a sort of subcontractor, arranging for placement of Chinese trainees in a series of mines and securing local housing.[17] As we shall see later in the chapter, training was not the only service Pierce provided for the Nationalist government.

Important to note is the role that Taub had played in this introduction. In 1944, Taub had worked as chief of the engineering service of the

FIG. 5.1. "China's Needs for Industrialization" from the Taub Plan for China's postwar industrialization, created by Alex Taub and a group of FEA engineers. Source: Shire, "Abstract, Guide to the Industrialization of China."

FEA. His team of twenty-five engineers, operating in the United States using data and information supplied to them by the NRC, China's Ministry of Communications, and numerous Chinese engineers in the United States, drafted a "Guide to the Industrialization of China" that became known as the Taub Plan (see fig. 5.1). Taub traveled to China with Nelson in the fall of 1944 and presented his plan to Weng Wenhao, who gave "valuable criticism." The final draft was made available to both the Nationalist government and select U.S. government officials in January 1945. As they prepared the plan, the FEA engineers sought assistance from "leading American industrial and engineering firms" as well as "leading American universities," and the U.S. Public Health Service.[18] The plan offered a set of macrolevel strategies for the modernization of Chinese industry, agriculture, and health care using U.S. "technical and economic assistance." As the abstract for the plan noted, "The most important contribution that can be made by the people of the United States to accelerate this development [of China] is to enlist our technical ability and knowledge."[19] From this cooperation, benefits would accrue to both China and the United States. For the United States, those benefits would be the development of new markets for American goods and new outlets for U.S. capital, an increase in industrial employment in U.S. factories that would supply machinery to China, and "the strengthening of economic ties between China and the United States over a long period through their cooperation during the initial stages of China's industrial development and specifically through the introduction of American standards, techniques, and industrial practices."[20] By participating in industrial training programs, companies like Pierce Management were fitting themselves neatly into this vision of postwar Chinese American cooperation.

THE TRAINEES

Most of the trainees who traveled to the United States in 1945 were selected and sent in groups according to their fields. For example, one group of 428 trainees arrived on September 3, 1945, with the expectation of remaining in the United States for one year. The training programs for these technicians encompassed a wide variety of subjects, but the largest numbers of trainees were in the fields of railway engineering and administration (ninety), highway engineering (eighty), aviation and navigation (seventy), and telecommunications (sixty). Another group of 117 trainees arrived a few weeks later on September 26. In that group, machinery manufacturing, chemical industries, mining, and electrical manufacturing were the areas with the largest numbers of trainees. This group expected to remain in the United States for two years of training.[21]

The ITA and the NRC identified trainees in terms of broad training categories such as industry and mining, agriculture, or communications, and within those categories, trainees belonged to subgroups, such as electrical power, coal mining, industrial management, or chemical engineering. Each trainee, however, had specific training goals, and it is clear from the monthly reports that trainees sent in to the ITA and the NRC that many of them took these goals quite seriously and were not just passively accepting whatever training the ITA provided them. In cases where placements were not working well, for example, trainees requested that the ITA find them a different placement. In cases where they came to realize that it would be beneficial to pursue an additional line of learning such as an alternative manufacturing process for a product, trainees asked the ITA to place them at factories where they could learn that process, and so on.

Trainees in some fields went in groups for longer-term training placements. Others were sent independently to placements that matched their individual training goals. Trainees typically remained in the United States for about a year, though some extended their stays for as long as several years. The first large group of trainees started returning to China in June 1946.[22] New groups continued to arrive at least through the end of 1947. The program officially continued until December 31, 1948, at which time the training department of the NRC-NY closed. From January 1, 1949, the Chinese government would only provide reduced services to those trainees who had requested to extend their training time in the United States.[23]

Trainees filed monthly reports with the ITA that were passed along to the NRC, and at the completion of their training, they also wrote up final reports that went to both the ITA and the NRC. Most of the reports were written in English. In general, the monthly reports provided detail on precisely what the trainees were doing with their time. Some included expansive descriptions of industrial processes and even technical drawings of the machinery that was used in those processes. The final reports varied in length and depth. Some were merely reiterations of the training schedule, but others showed that trainees were thinking deeply about how to apply in China what they had learned in the United States.

One example of a trainee who put all the pieces together into a comprehensive final report that not only described what he had learned but also considered its relevance to China and formulated a plan for the future was Jiang Shanxiang 江善襄. Jiang was studying the phosphorus industry as part of the industry and mining group, an industry with which he already had considerable experience in China. He trained in ten different plants in the United States, investigating several distinct manufacturing processes. Jiang was specifically interested in the machinery that was used in these processes and studied the comprehensive blueprints of every plant he visited. He sketched more than sixty to-scale drawings of machinery with dimensions noted, wrote up flowcharts that described the entire industrial process in each plant, and collected information about equipment prices. Jiang clearly took his work quite seriously, recognizing that "the fertilizer industries are very important. If we want to become a strong nation, if we want to have strong bodies, the fundamental requirement for these is to have enough and better quality food. The total farming land to the total population of our country is less than U.S. With the limited land with the fixed labor, the only way to produce more and better food is to applied [*sic*] fertilizers."[24] Jiang clearly saw his training and his profession as being of considerable national importance, but his time in the United States also convinced him of just how much more important the manufacture and broad application of fertilizer might be in China even than in the United States.

Jiang returned home with a clear set of ideas about how best to put his training to use in China and having developed thoughtful plans for doing so. In his final report, he proposed a practical three-step process for building a Chinese phosphate fertilizer industry that took into account all that he had learned in light of the conditions in China. First, he suggested the construction of an "ordinary superphosphate plant with sulphuric acid plant, because this is most reliable and easy to build." Along with the construction of the plant, he noted, the Ministry of Agriculture and Forestry (MOAF) would need to undertake extension work to demonstrate the use of manufactured fertilizers and thereby increase their use by farmers. The second and third steps to his plan were to build increasingly complex plants that he felt produced better products. The third stage, he noted, could not be undertaken until there was sufficient electricity to support the plant because it relied on electric furnaces. Jiang was not just a dreamer, though. He wanted to get the project going as soon as possible. "I hope," he wrote at the conclusion of his final report, "we can start the first step at this year."[25]

Liang Shen 梁燊, of the MEA and a member of the industrial management group, also utilized what he had learned to develop a holistic plan for the development of his industry. Liang was assigned to the automotive industry with a focus on commercial trucks and spent his training period at International Harvester and White Motor Company building a familiarity with the "functional organization of the industry with special emphasis on methods of planning, processing, production scheduling, material control, time-study and job-evaluation." He was particularly impressed by the well-ordered interaction between the various divisions within the factories and also by the relationships that had developed between factories and parts suppliers. He observed that "the development of the automotive industry in the United States has gone to an extent where most of the technical difficulties are solved. Almost every part has a special machine and tools for its productions." The automotive industry in China, however, had a long way to go to reach that state. From Liang's perspective, the most logical approach to take in China would be to start with the construction of an assembly plant that would, at least initially, utilize imported parts. By assembling trucks in China instead of purchasing them fully made in the United States and shipping them to China, Liang estimated that China could save $500 per truck, funds that could be used to build machine shops, a foundry, a forge shop, and heat-treating departments at the site of the assembly plant. All of these things would be required for China to be able to manufacture its own trucks. A second reason for taking this graduated approach, Liang argued, was that China's iron and steel industries were not sufficiently mature in 1946 "to meet the quantitative and qualitative requirements of the automotive project."[26] Liang's training had helped him to see the big picture and to map out all of the steps in the automotive manufacture process. In doing so, he was able to see the gaps in what China could bring to such a project as of 1946 and come up with a pragmatic plan that would provide solutions for an immediate problem while buying time for the development of basic industry (iron and steel) that was required for advanced manufacturing, as well as for the construction of facilities in which all the manufacturing stages of his industry could be developed.

In general, trainees reported being impressed by both the way American plants operated and the people who operated them. Guo Bingyu 郭炳瑜, who studied the alkali industry as part of the industry and mining group, reported being "deeply impressed by the good organization and high efficiency in the American plants. Their scientific control methods with continuously improve [*sic*] are far ahead from China. At the same time, all the plants are going to maintain highly safety record [*sic*]." He also felt that it would be essential for China to emulate this. "As my own opinion," he wrote, "unless we put these points in our industry, we can't compete with other countries."[27] Interestingly, Guo was not just thinking about building a domestic economy; he was also imagining Chinese industries that would be internationally competitive. But to reach that goal, Chinese industry would have to adopt the patterns of organization, efficiency, and safety that he had observed in the United States.

Huang Yi 黄翼, who had studied industrial management with a focus on accounting procedures at the TVA, the Hudson Coal Company, and Columbia University, reported that "I believe that the most important factor of highly developed industrialization in the United States is not the technology but the men who operate and manage the industry" and went on to say that he felt that it was "very important for us to study the spirit behind the gigantic industrial system." However, he also noted that there were other lessons that China could learn from a study of the industrialization process in the United States. "As a management man from China," he wrote, "I was not only interested in the merits and successes in the industrialization of the United States, but also in the deficits and mistakes made in the development—such as labor troubles. For shortening the time of industrialization in China, it is important to avoid the mistakes as far as we can."[28] China could also learn from and avoid the mistakes and problems of American industry.

Although the NRC took considerable care to train people in a wide range of areas of need, and trainees, as noted above, had specific goals for their training, it was not always completely clear either when the training was being set up or as trainees were heading home, exactly what they would do with their training upon return to China. The situation in China was changing too rapidly for certainty. As a major in the Chinese army requesting training for one of his men wrote to S. C. Wang of the NRC-NY and CSC, "As you know, our officers are being trained for Army technicians. Even though the war is ended, our program is still being carried forward. The men trained will either work for the Army or for post-war industrialization."[29] Training was intended to support larger plans, but political and economic instabilities in China meant that even those who sent people for training were not always sure what the trainees would be doing with their new knowledge after they returned to China.

A second problem was the question of just how easily the learning that trainees were doing in the United States could be applied to the Chinese context. Whereas Chinese trainees in the United States were typically both impressed and inspired by the American patterns of industrial management that they encountered during their training periods, they nonetheless tended to believe that the export of such patterns to China could not be done wholesale. Liu Shoudao 劉守道, an industrial management trainee and employee of the NRC who had been sent in the first large group of trainees in June 1945, observed that "I cannot tell now exactly how much I have benefited from my training, but I can be quite sure that I have done my utmost and wasted no time during my one year stay in the United States."[30] Liu had wholeheartedly engaged in the training enterprise but felt strongly that American know-how could not simply be transported to China without regard to context. "It is my personal opinion," he wrote, "the most important thing for us to learn in this country is the main points of their systems, and their methods and spirits of doing things. Circumstances in China are not quite the same as that in the United States, we can not copy their whole system without any change. As things that are all right in this country are not always all right in other place, we should, when returned, work out our own systems to adapt our own circumstances, but use what we have learned here as an important reference or even a basis." Liu had put his finger on a very important point for all of the trainees. While in the United States, they were being exposed to American industrial norms and methods. But as we have already seen in chapter 4, American ways could not be exported wholesale to China because the circumstances were different. The trainees would have to adapt U.S. know-how to local conditions, and that would require the kind of thinking about doing things in stages and adapting to the limitations of existing resources and technologies that Liang had suggested with his truck plan.

Liu also believed strongly that the first step toward development in China was to build modern industry and harness natural resources. To do that, he felt, required a body of appropriate personnel, "especially technicians." In the conclusion of his final report, he urged the Chinese and American governments to continue the training program but to be mindful of how the training fit into China's overall development program. "Maybe our government will send more trainees, one group after another, to this country for training," he wrote, "but it must be beared [*sic*] in mind that this kind of training as any others should be carried on under a whole plan and with a well-prearrangement. Otherwise the result will turn out not as satisfactory as desired."[31] Training for training's sake, Liu felt, was not a good use of resources. Echoing the sentiments of many of the planners described in chapter 1, Liu argued that China's need for trained technicians was substantial, but his own personal experiences led him to believe that training needed to be carefully planned to fit China's particular needs.

But how could planning be effectively matched up with the realities of a rapidly changing situation on the ground in China? Many trainees returned to China having acquired a great deal of potentially useful knowledge that they hoped to apply to Chinese industry, mining, and agriculture, but training programs lasted for a year or longer, and political and economic conditions in China were changing constantly. As a result, the question of what they would do upon their return to China became increasingly murky over time. Like a number of the consulting engineers the NRC hired to assist with specific projects (see below), the trainees were often being prepared to work in industries over which the Nationalist government had no control owing to territorial losses in the civil war.

For example, Manchuria had appeared at the end of World War II to be a land of considerable economic opportunity, not least because it was filled with natural resources and the Japanese had developed its industrial economy over the previous fifteen years. Despite a successful early Nationalist military campaign in southern Manchuria, however, Russian occupation of some parts of Manchuria, Chinese Communist occupation of others, and a virtual stalemate between the three parties for the first year or two after the end of the war meant that the political situation was simply not stable enough to permit implementation of Nationalist plans for Manchuria's economic development.[32]

In sum, the large-scale postwar training program went a long way toward building a body of technical experts who ought to have been able to oversee nearly every dimension of China's postwar industrialization. Trainees returned to China with hands-on technical experience and both theoretical and practical knowledge of processes and machinery used in the manufacture of numerous critical products, plant organization and operations, industrial labor management, industrial accounting, and so on. Many of them clearly indicated as they returned home that they understood that American systems and processes could not be transported wholesale to China, where the industrial infrastructure was not nearly as developed. As a result, they anticipated, and in some cases explicitly planned for, the need to adapt what they had learned to suit the Chinese context. The sources reveal little, however, about the extent to which any of these trainees actually managed to implement their training following their return, and for those who did, whether they continued to do so long term in China, or if they migrated, along with their skills, to Taiwan or elsewhere.

Changes in Postwar Aid Structure and Institutions

As we have seen in chapters 2–4, both the Chinese and U.S. governments, aided, in some cases, by nonstate actors like the China Institute, developed a number of models of technical training and consulting during the war. This included training programs in China and in the United States as well as structures through which general technical advice could be sought and given. As the war ended, programs that had been, on the U.S. side, managed by government entities such as the State Department, the White House, or the FEA, drew to a close. Even so, they formed the foundation for and to some extent served as models for continued collaboration between the United States and China. During the war, most U.S. technical advising and training in China was undertaken by experts sent to China through the State Department's cultural relations program or by technical consultants associated with the American War Production Mission (AWPM), both of which were programs managed by the U.S. government, although the NRC did also procure some technical advice directly from U.S. firms through the UTC. By early 1945, the policy of the U.S. government was to aim for "the most rapid possible return to private trade with China" in the postwar period.[33] The U.S. government was not looking to disengage with China; it just wanted the work of economic engagement and postwar rebuilding of China to be taken over by the private sector.

Certainly, the ITA training program, the AWPM, the FEA planning work that had resulted in the Taub Plan, and even the cultural relations program had positioned many U.S. companies to develop their own

direct relationships with Chinese government agencies or with other actors in China, but as the U.S. government stepped aside at the end of the
war, there was no longer any U.S. entity to coordinate the relationship.
With the creation of the Chinese War Production Board (CWPB) and
the AWPM, the U.S. government sought to control, or at least guide,
Chinese industrial, and to a certain extent, agricultural development and
expected the Nationalist government to follow the advice it was given.
But as the end of the war drew near, the NRC, still welcoming whatever
assistance it could get but at the same time pushing back against these
efforts at control, began to take an increasingly leading role in building
(or, in some cases, expanding on) relationships with private U.S. companies, contracting with them to undertake both training and rehabilitation work.

The U.S. government sponsored wartime technical advising and
training model in China had depended largely on the initiative of the individual technicians who had been sent by the United States. These
technicians were given broad remits and access to the relevant industries,
research stations, and other sites by Chinese government agencies such
as the NRC and the MOAF. Technical advisers in these programs were
typically there in response to specific requests or concerns that had been
identified as important by the Chinese or U.S. government; however, they
were also generally encouraged to offer suggestions for whatever they saw
that they felt could be fixed. A number of them, such as Dykstra, also
understood the creation of training programs to be part of their remit,
and they developed such programs to the extent that they felt able to
do so.

The wartime programs were funded, for the most part, by the U.S.
government, and as such, they were constrained and shaped by the opinions of U.S. diplomats and technical advisers and by the nature of U.S.
interests. The U.S. embassy in Chongqing did not, for example, endorse
every request for a technical consultant that the Nationalist government
put forth.[34] In parallel to these efforts, the Nationalist government developed the infrastructure in the United States in the form of the NRC-
NY and the UTC, and relationships with State Department and AWPM
advisers, that permitted it to reach out more directly to more U.S. industries that it had been in contact with before the war. Many of these connections were aimed solely at procurement of industrial parts and ma-

chinery, but some, as Ying Jia Tan has shown, also facilitated the transfer of knowledge.[35] These institutions and relationships allowed China to take greater ownership of the technical advising process and to get around some of the constraints imposed by U.S. diplomats, the FEA, the AWPM, and other such entities. Even before the end of the war, the NRC-NY was making direct connections with potential U.S. consultants (such as Pierce Management, already mentioned, and Arthur G. McKee, described below) to set up rehabilitation contracts for certain industries, and as the war drew to a close, that process intensified. Such activities were undertaken with the knowledge of U.S. government officials and comported with the U.S. policy of shifting the work of postwar rebuilding to the private sector, but they did not have to be arranged through State Department or White House channels.

In theory, the relationships established through these contracts provided a greater level of autonomy for the Chinese government, and the enthusiasm with which the NRC went about making contracts in the late summer and fall of 1945 suggests that NRC leaders felt they had considerable potential as a mechanism for soliciting targeted technical assistance to specific industries. Financing the contracts, however, was tricky. The Nationalist government had little revenue that could be applied toward such activities and, particularly in the context of the last months of the war, expected that lend-lease funds could be applied to rehabilitation projects. There is no doubt that the Nationalist government was hoping to develop a program through which they would have autonomy in the selection of contractors, but wherein the services those contractors would provide would be supported to a large extent by American money. With the end of the lend-lease program in early September 1945 and of the FEA in December 1945, however, U.S. money for technical projects started to dry up. Although a great deal of U.S. aid continued to make its way into China after the war via mechanisms such as the United Nations Relief and Rehabilitation Administration (UNRRA) or the State Department, most of that aid was either earmarked for military use or granted in the form of foodstuffs and other items to be distributed for relief. To a limited extent, that aid opened up resources that the Chinese government could redirect toward rehabilitation projects, but to really undertake the kinds of massive reconstruction projects the NRC and its technical advisers had in mind, the kinds of projects anticipated in the Central Planning

Bureau's Five-Year Plan for China's Economic Development, would require substantial loans from the United States. At the same time, management of postwar U.S. loans to China (at least loans that were focused on development) shifted primarily to the Export-Import Bank, which had granted loans to China in the past but was fairly conservative in its approach.[36] This shift led to the creation of a new set of procedures through which China could secure loans. On the Chinese side, however, evidence suggests that the requirements of the new system were not entirely clear, and it took the Chinese government time to figure out how to navigate this new lending system. For example, in the case of Pierce Management, the NRC-NY had a good deal of trouble getting reimbursed for funds they had paid out to Pierce for a coal mining report that company was producing.

Just how U.S. technical aid to China should be managed was, in fact, a subject of some debate in the United States in the months following the war. Although some advisers continued to advocate for and support the Chinese government's continuing efforts to develop and rehabilitate industry, there was skepticism in some circles about whether the Chinese could adequately manage a large influx of rehabilitation aid. That skepticism was most evident among the people with the actual power to provide that aid, but even Weng Wenhao worried that China would best manage U.S. aid with the assistance of U.S. advisers.

Edwin Locke and others in the U.S. government remained in contact with Chinese political leaders and NRC staff and continued to offer encouragement as well as assistance in facilitating the development of relationships, but they ceased to play the same kind of guiding role that they had performed through the AWPM in the early part of 1945. By early 1946, Locke had been reassigned by President Harry Truman to take on new tasks, but he nonetheless continued to serve, at the invitation of the Chinese government, as an economic adviser to the national government of China. In this capacity, he indicated to Weng Wenhao, "I sincerely hope that you and your colleagues will continue to call on me at any time that you feel I can be useful," and he appears to have meant it. Locke kept tabs on the consultants who were going to China and requested and received reports on them from the NRC-NY. He wrote to Weng that he had been meeting with many of these experts as they left for China and that he had "done all I can to give them a thorough and constructive approach to China's problems."[37] Locke kept up his interest in China for some time and continued to advocate for U.S. assistance to China well after the end of the war, though not always the full range of assistance that China was hoping for. As he wrote in a memo to President Truman in December 1946, "What, in my opinion, the Chinese need above all is American aid and guidance in helping them to do what I am convinced they want to do but don't know how to do. Large loans from us are not to my mind a decisive matter at this time, but instead China requires help and know-how in getting much more out of what she now has."[38]

Much as Nelson had insisted on the creation of the CWPB, Locke was also insistent that China needed to develop its own central structure to oversee postwar economic development. His vision was for a postwar replacement for the CWPB, and he called for a new institution that would both plan for and manage China's postwar rehabilitation. Understanding that, once again, U.S. aid might be contingent on the existence of a central economic institution, Chiang Kaishek and the Nationalist government leadership established the Supreme Economic Council (SEC, 最高經濟委員會) in November 1946. It existed until May 26, 1947, when it was reorganized into the National Economic Council (全國經濟委員會). The SEC was jointly headed by T. V. Soong and Weng Wenhao, and its members included heads or highly placed personnel from major ministries and other government entities, including the Ministry of Education, the NRC, and UNRRA in China. At its inaugural meeting on November 26, 1945, Chiang Kaishek declared that the SEC would oversee the revitalization of industry, the construction of transportation infrastructure, the modernization of agriculture, reform of both rents and taxation, and elimination of government corruption. It would accomplish these goals by coordinating the work of all government ministries that had influence over economic matters and by making "clear-cut and rapid decisions on economic policy."[39] In other words, the SEC would behave much as the CWPB had been intended to act. In fact, among the first acts of the new entity was the abolition of the CWPB, the utility of which had run its course. It took some time, however, for the SEC to begin functioning in any meaningful way. As Weng Wenhao wrote to Edwin Locke a month after the SEC's inauguration, the entity had been created, but "no further action has yet been taken." Weng opined that the reason for this was that there had been too many pressing things to take care of to be able to execute the kinds of long-range or macrolevel plans that the SEC was intended to oversee. "There have been made general plans for constructing railroads, new industries, water conservancy work, etc.," he wrote. "Just in this category of work, it is very necessary to secure true coordination and better judgment. China is entering a new stage of her national life. The right direction and central planning is indispensable to the establishment of a right channel for the whole nation to go through. It is therefore very necessary that such new institutions as the Supreme Economic Council be provided with competent staff and good advisers in order to lay a sound foundation for our national economy." Weng saw the value in having a central entity to guide China's economic development. However, he also felt it was important to staff it well. To that end, he was eager for American assistance in providing advisers who could help the SEC, as Locke and others had helped the CWPB, and he made sure to mention in his letter that he hoped Locke had conveyed this wish to President Truman.[40] Locke's goal of seeing China's government develop a central infrastructure to oversee economic reconstruction and development appears to have been only partially fulfilled. The shell was created to much fanfare, and at least some of China's major economic leaders understood the potential importance of such a unit, but Weng, at least, was not ready to put the whole thing into Chinese hands. He clearly felt that China still needed American advisers to help right the ship.

Weng was not the only person to have these concerns. In April 1946, Seymour J. Janow, an economist who had worked for the FEA and had also spent time in China as an economic consultant to Claire Chenault, gave Locke a confidential memo he had written for the Export-Import Bank titled "American Technical and Administrative Personnel for the Chinese" in which he articulated a number of concerns. In particular, Janow was worried that China was experiencing a large capital inflow and taking on management of new properties in Manchuria and Taiwan all at the same time and that this expansion of responsibility was likely to be more than the nation could handle. "Not only do the Chinese have the enormous new capital to absorb by way of repairs and rehabilitation of pre-existing industry and new development," he wrote, "but they have the coincident problem of undertaking expanded enterprises. . . . Clearly, China's resources of administrative and technical and planning personnel are not sufficient to solve these problems efficiently." Janow felt that for the sake of both the successful development of Chinese industry and the development of a sound pattern of Chinese-American economic relations, China should employ a principal economic adviser, perhaps General George C. Marshall, Mr. John Winant, or Lt. General Albert Wedemeyer, who could "recruit for the Chinese the personnel needed and draw when necessary on many companies and from all fields of endeavor. All advisors would be employed by the Chinese, thus avoiding any danger that the Chinese would feel Americans were intervening in their internal affairs. This scheme would also avoid the dangers of special American monopolies in the China trade or in new investment." Janow was, in essence, suggesting a new version of the AWPM because he believed that the Chinese government was simply not competent enough or in command of a sufficient number of technicians to oversee China's redevelopment. At the same time, he made it clear that the new structure should be independent and should not be a U.S. corporation, which was the direction the Export-Import Bank had been considering heading in. Janow was particularly concerned that some system of advisers be set up before reparations materials from Japan and capital equipment from the United States began to arrive in China so that these materials and equipment would be put to the best possible use.[41]

From the perspective of the Export-Import Bank, which was managing the line of credit that the United States had extended to China, the creation of the SEC seems not to have been as important as the peace talks that took place in the winter of 1945–1946 as a determinant of whether they would extend rehabilitation loans to China. Although the Nationalist government started submitting loan requests to the Export-Import Bank very soon after the end of the war, the bank was essentially unresponsive. Its refusal to grant loans was based on the belief that there was no logic in funneling large sums of money into a nation that was neither politically nor economically stable. In the first quarter of 1946, however, the situation in China had stabilized enough that the bank started making some loans and eventually authorized $82.8 million in loans to China. These loans went primarily to transportation and power projects as well as to cotton development. In addition, one was granted to Yongli Chemical Industry, a public-private partnership that was engaged in fertilizer production and had ties to the cotton industry but that had also, at least during the war, processed tung oil for use as a fuel. In April 1946, the

Export-Import Bank authorized an earmark of $500 million to be granted before June 30, 1947, but never disbursed any of the funds even though it was encouraged to do so by both General Marshall and the State Department.[42] Frustrated with the bank's inaction, Marshall and the State Department sought, in late 1947, to encourage the U.S. Congress to appropriate a sizable fund to support China.[43] In February 1948, President Truman also threw his weight behind the project. By March 1948, Congress had decided to act on this request, passing the China Aid Act, which authorized a substantial aid package for China, $67.5 million of which would be earmarked for reconstruction projects focused on electrical power, transportation, and coal mine rehabilitation.[44] In December 1948, however, all reconstruction projects were halted at the request of the Economic Cooperation Act Administration, which was overseeing the loan. Concrete U.S. government support for Chinese development projects, therefore, declined after the war even though the potential for such support was consistently there.

Postwar Rehabilitation Contracts

Oversight of U.S.-funded projects and of the rehabilitation process in general was not the only concern held by American advisers to China. A great deal of their focus, particularly in the immediate postwar period, was on personnel who could implement modernization projects at the level of the specific industry and even plant. As Donald Nelson, anticipating the imminent return of Japanese-controlled territories to the control of the Nationalist government, wrote to T. V. Soong just after the end of the war:

> It is of the utmost importance that the large industrial facilities which China is taking over from the Japanese in Manchuria, Formosa, and the Eastern provinces be put into peacetime production as soon as possible. Since it is unthinkable to permit the Japanese to continue to manage these plants, and since plants which are shut down for any length of time deteriorate very rapidly, new and qualified management obviously has to be provided at once. My personal observation in China makes me feel that at

the present time your country does not have enough experienced industrial managers for this purpose. A considerable number of highly trained managers will be required to reconvert these big Japanese-built plants to the making of peacetime products for China and to run the plants efficiently thereafter. The best course which your government could follow, in my opinion, would be to complete arrangements for temporary management of these plants at the earliest possible moment with American companies capable of doing the job and of training Chinese managers."[45]

Locke, writing to Truman a few months later about the newly established SEC and the economic program it proposed to oversee, echoed Nelson's concerns, noting that "left to their own devices, the Chinese will find it exceedingly difficult if not impossible to carry out many important parts of this new economic program. China lacks the administrative and managerial experience and personnel necessary to get such a program soundly under way, not to mention shortages of essential materials and physical equipment."[46] In fact, many of the people who might have been able to serve in such management positions were at that very moment either in the United States or preparing to travel to the United States as part of the large-scale training program mentioned above. That program was only just getting under way as the war ended, however, and would not begin to yield appropriately trained personnel until nearly a year later.

The NRC was in full agreement with Nelson and Locke's assessment and, to help solve its management personnel problem, began to expand the number of U.S. firms it was contracting with for consulting services immediately after the end of the war. Reporting in April 1946 to Edwin Locke, the NRC-NY wrote that "in order to hasten the industrialization of our country, we have found it necessary, due to the lack of inexperienced [*sic*] industrial management, to seek aid from foreign countries (such as the United States), where the various key industries have reached a peak of industrial development. Thus, by obtaining the guidance of their expert engineers in the respective fields and by studying the most modern methods, we shall soon be able to utilize our vast wealth of undeveloped natural resources, fully employ our large amount of labor, stabilize our national economy and restore our country's wealth. With this goal in mind we have negotiated, and are still negotiating, with prominent firms in the United States for their technical cooperation."[47] The firms with which the NRC contracted could assist with both rehabilitation in the present and training for the future.

In the eight months following the end of the war, the NRC signed re-habilitation contracts with Pierce Management for assistance with rehabilitation and modernization of coal mines, J. G. White Engineering Corporation for help with power facilities in Manchuria and elsewhere in China including Taiwan, the Chemical Construction Corporation for help with the restoration of chemical plants in Manchuria and Taiwan, Universal Oil Products Company for help with rehabilitation of oil refineries and coal hydrogenation plants, General American Transportation Corporation for assistance with development of cement plants in Manchuria and Taiwan, Arthur G. McKee Company for help with iron and steel works in Manchuria as well as north and central China, E. A. Rose Company for assistance with sugar production and refinery, the American Cyanamid Company for help with calcium cyanamide plants, the H. K. Ferguson Company for help with electrolytic alkali chlorine plants, and the United Engineering Corporation for recommendations regarding oil and gas production. As of April 1946, engineers or technicians from each of these companies had traveled to China to survey the situation and write up reports with recommendations.[48] By June 1946, even more American technicians from other companies had been brought to China by the NRC to engage in similar work in their respective fields. These included staff from the Fitchburg Paper Company to rehabilitate the paper industry in Taiwan and Behre Dolbear & Company to help with tungsten, antimony, and tin mines.[49]

These rehabilitation contracts had several overarching aims. The primary goal of the contracts was to bring American know-how to bear on Chinese technical problems so that the American consultants would provide concrete advice on how to develop or rehabilitate specific Chinese industries. As a secondary matter, like Nelson and Locke, the NRC believed that China lacked adequate managerial talent and anticipated that in at least some cases, the companies with which they were contracting would temporarily manage Chinese industries and train Chinese personnel to take over following that interim period. It is also clear, however, that even in spite of the fact that the contracts were with foreign firms, the agenda of both the NRC and the contractors was very much to get Chinese industries running under Chinese management and under the authority and control of the Nationalist government.

A third possible motivation for these contracts, and one that was suspected by some of the American contractors, was simply to put some window dressing on Chinese development efforts in an attempt to get more U.S. aid. W. S. Finlay, executive vice-president of J. G. White Engineering Corporation, reported to Edwin Locke's assistant, Arthur Lowery, that J. G. White's "engineers believe the major reason the Chinese encourage these technical missions and consultations is to give a business-like appearance which they hope will lead to continuing loans."[50] Evidence suggests, however, that these contracts were not mere window dressing. The Chinese government genuinely needed assistance with redevelopment, and it used the contracts to get ideas and develop connections that could be fruitful even in cases where the NRC did not opt (or was unable) to pursue all of the follow-on development strategies proposed by the companies. Primarily because of limited financial resources, it was not uncommon for the NRC to pursue only parts of the proposed reconstruction projects and to do so with the resources it could secure from the Bank of China. So whereas the consultants' reconstruction reports often proposed a step-by-step plan for the redevelopment of an entire industry across the country, the NRC could usually only muster the resources to undertake one or two of those steps, had to do so on the cheap, and continued the relationship with the consultant primarily for the purpose of purchasing materials.

The NRC rushed to enter into a variety of rehabilitation contracts in the last months of the war and immediately after its end and was able to honor some of those contracts in late 1945 and through 1946. Those were the contracts that aimed at production of an overarching reconstruction plan for a particular industry. The NRC also negotiated other contracts for more specific activities such as the takeover of particular plants or to supply specific industrial machinery, but successful execution of this kind of contract was dependent on Nationalist control of the territories in which the work would be conducted and availability of funds. In spite of the fact that U.S. loans for rehabilitation projects were only rarely forthcoming throughout the immediate postwar period, the Chinese government continued to submit loan applications and U.S. firms continued to provide letters of support in hopes of getting the business. In some cases, however, the NRC was compelled to postpone the execution of a contract. In one example, a contract was canceled with E. B. Badger and Sons

Company for engineering services to help rehabilitate chemical plants in Manchuria, north China, and Taiwan that had been negotiated in the summer of 1945 and signed on September 20, 1945. By May 1947, both sides had determined that circumstances beyond their control had made fulfillment of the contract impossible, and they agreed to cancel it.[51]

The companies with which the NRC entered into these contracts were by and large self-interested. The initial contracts were typically for the company to conduct a survey of an industry and develop a plan for that industry's future development, which the NRC would then use as the foundation for an application to a loan-granting entity such as the Export-Import Bank in Washington, DC. The plans these industrial advisers developed were never cheap, nearly always running in the tens of millions of dollars no matter what the industry, and the advisers generally offered their own services in the execution of the plan. In other words, if American money was going to help finance Chinese reconstruction, the contractors certainly believed that much of that money should flow back into American hands. To that end, observers such as W. S. Finlay of J. G. White Engineering felt that there should be a role for American consultants in overseeing or guiding any development work that would be undertaken with U.S. assistance. As he indicated to Lowery, "In connection with any financial aid granted China the agreement must specify setting up a high-caliber economic and industrial organization on a continuing basis to supervise administration of the funds, this organization to be staffed with American personnel."[52] Interestingly, J. G. White did go on to position itself to take on just such a role with regard to the rehabilitation of Taiwan after 1949.[53]

CASE STUDY I: IRON AND STEEL

In the early spring of 1945, the NRC began negotiating a contract with Arthur G. McKee and Company of Cleveland, Ohio to consult on the rehabilitation and improvement of iron and steel works in China. The March 14, 1945, contract tasked McKee to send three engineers to inspect and study iron and steel works in China, prepare recommendations and plans for existing plants, examine feasibility of building additional plants, and examine additional data, all of which would culminate in an "all China" plan for iron and steel development over the following five to ten years. This final plan would "embody recommendations of McKee for the manufacture of products among the different plants, the sequences of construction of such plants and the selection of regions of location for such plants."[54] At the same time, the contract asked McKee to identify a single integrated plant in southwest China for which the company would produce a detailed report and a plan for its development as the first big step toward modernizing the iron and steel industry. In October 1945, the NRC asked McKee to replace "southwestern China" with "mid-Yangzi region," noting that "as the war is now ended, all necessary field studies are possible for other projects also. From the standpoints of transportation, raw materials and market, a plant in the Mid-Yangzi region should be studied first."[55]

That October, the NRC negotiated a second contract with McKee. This contract was also for surveying and planning work, but this time, specifically in Manchuria. It was different from the first contract, however, in that it also arranged for McKee to help directly with the take-over of plants and the training of personnel both in China and in the United States for those plants. The Manchuria contract, which was signed in late October 1945, also stipulated that the NRC could hire McKee's engineers to remain in China running NRC plants after the five-month term of the contract expired.[56] Precisely why the NRC elected to work with McKee rather than one of the bigger steel companies in the United States is unclear, except for the fact that there was an existing relationship to build on. William A. Haven, vice-president of the company, had spent time in China in 1933 and 1934 consulting on blast furnace and steel plant construction, and it seems likely that this shared history led the NRC to return to McKee for help more than a decade later.[57]

It took some time for the activities outlined in the first contract to be realized, and owing to conditions in Manchuria, the second contract was never carried out. After some delay, Haven, along with Kenneth A. Barren and Earl H. Collester, traveled to China from March to June 1946 to tour sites in and around Nanjing, Beijing, and Chongqing, collect data, and develop a plan for development of the industry. The NRC supplied them with ample data prior to their travel from which they produced a preliminary report even before they reached China. Once in China, they inspected existing facilities as well as possible new locations, taking into account proximity and quality of raw material reserves and transportation options for each site. In addition, they met with NRC engineers, railroad officials, and other government officials, who supplied them with an array of data, and they studied market conditions as well as industrial development plans with an eye to determining how to prioritize the goods that should be produced. As the McKee report (the one produced after the site visits) noted, "It is obvious that you will ultimately wish to provide facilities for the production of wire and wire products and pipes, tubes and fittings and it is desirable that you do so in order to provide more diversification of marketable products. However the market for these particular products, in sufficient quantities to justify expenditure of the money required to provide facilities for their production, does not exist at the moment and it is not likely to be created in the next few years. On the other hand, the market for rails, plates, sheet and tin plate, structurals and bars to take care of China's immediate and near future requirements, and these requirements are urgent if China is to progress industrially, would seem to require production of these items in the first plants to be built."[58] McKee's evaluation was thus shaped by a fairly comprehensive consideration of the current conditions of both industry and transportation in China and the ways in which the iron and steel industry could best support overall industrial development.

The McKee report called for bringing plants at Shijingshan (located in the Beijing district), Chongqing, and Daye (located near Wuhan) on line first. The plants would produce such things as rails, fittings, plates, and large bars, all of which would be useful for developing railways, and other materials that could be employed in munitions plants and other factories that needed to improve their machinery to manufacture smaller-scale goods with a considerably smaller market. The Shijingshan plant was in the best shape, and for that reason, the McKee engineers proposed rehabilitating it first. The Daye plant was selected as the "integrated plant" that the NRC had requested McKee identify in the middle Yangzi valley and would be entirely rebuilt with the intent that it would ultimately become "the keystone of the future iron and steel industry." For that reason, the McKee engineers proposed that no "second-hand or obsolete equipment" be used at that site and urged the NRC not to "omit anything in its initial construction that is essential to its further expansion and to the attainment of minimum operating costs."[59]

For various reasons (most notably poor infrastructure and high cost of transportation and thus production), the McKee engineers did not favor prioritization of the Chongqing plant. However, NRC officials, in particular Qian Changzhao 錢昌照, now heading the NRC, were insistent that the Chongqing plant be given a high priority, owing primarily to its proximity to Nationalist-controlled munitions plants.[60] As McKee's Chongqing report noted, the arsenals and munitions factories located in the Chongqing district "represent large investments in time and money," and "to maintain and improve the supply of semi-finished iron and steel for these ordnance establishments, is a requisite for national defense, a consideration which the Government, according to our information, has rated above that of low production costs."[61] In other words, the NRC's insistence on prioritizing the Chongqing plant was motivated not only by strategic and political concerns but also by a desire not to entirely abandon the industrial infrastructure that had been built up in that area during the war. NRC leaders were surely also thinking about the ongoing importance of a Chongqing plant to continued efforts to develop China's western provinces in the postwar period. So even as the NRC directed McKee to look to the central Yangzi region rather than the southwest for an integrated plant site, NRC leaders still wanted to retain a major site in Chongqing.

Although McKee engineers did not end up undertaking any of the administrative or training duties proposed in the Manchuria contract, they did continue to provide services to the NRC at least through November 1948 and probably longer. Records show that McKee supplied drawings of plant layouts and material flow diagrams for both five- and ten-year plans for a number of sites including Daye, Chongqing, Beijing, and Canton, and they provided equipment and installation designs for Shijingshan.[62] As T. Y. Yen, vice-director of the Iron and Steel Administration in Nanjing, wrote to Sandys Bao of the NRC-NY, "To save foreign exchange," certain items being used in the reconstruction work at Daye were being sourced in China, but the NRC still needed "detailed drawings . . . for manufacturing, constructing or assembling."[63] The Daye project was not, however, completed before the Nationalists lost control of the Wuhan area in the civil war.[64]

One thing that is clear from this case is that the NRC was actively involved in guiding the direction McKee went in. Not only did NRC officials gather and supply the data that shaped the McKee team's preliminary understanding of the situation (though there is no reason to believe that the data were biased or flawed), they also provided guidelines for the kind of report they wanted, made their preferences for location of at least one of the key sites clear, and followed up with questions and additional requests. At the same time, the contractor provided a comprehensive and detailed report that could serve as a real and useful guide to the NRC as it determined how to proceed. As it turned out, the McKee plan could only be partly carried out owing to the larger political and military context, but the NRC did continue to work with McKee, even in spite of financial obstacles, to implement the parts of the plan that they believed they had a realistic chance of putting into effect.

CASE STUDY 2: SUGAR

As we have already seen, the NRC demonstrated considerable optimism regarding potential redevelopment of Manchuria and Taiwan. And as we have seen in the case of the McKee contracts, the NRC's aspirations for taking over enterprises in Manchuria did not work out. This made Taiwan all the more important as a potential source of both products and revenues for the Chinese mainland. As part of the rehabilitation contract system, American industrial consultants traveled to Taiwan to assess a number of industries, including petroleum and sugar.

Colonel Small, who undertook investigations of oil basins in Sichuan, Gansu, and Taiwan under the Hoover, Curtice and Ruby oil and gas rehabilitation contract, concluded that the NRC should develop Taiwan first "because of the available subsurface information obtained from the Japanese operations, the many installations of power, pipe lines and docks, and the good transportation facilities."[65] For the same kinds of reasons, the NRC was enthusiastic about redeveloping Taiwan's sugar industry and was already asking the AWPM to help find a sugar expert and to secure technical books related to sugar in late August 1945.[66] Sugar had been a booming enterprise under Japanese rule, but the industrial infrastructure had been seriously damaged as a result of bombing of both raw sugar processing factories and sugar company railroads. Nonetheless, the opportunity to integrate Taiwan's sugar industry into China's overall economy looked too good to pass up. Before the war, China had been a net sugar importer. Redevelopment of the sugar industry thus had the potential to replace imports and reduce outward flow of foreign exchange and perhaps even eventually expand production to the point that sugar could become an export item.

To this end, the NRC began negotiating with Arthur G. Keller, a sugar expert at Louisiana State University (LSU), in the early fall of 1945, soon after the Chinese government gained control of Taiwan. Keller initially indicated a willingness to travel to Taiwan to conduct a survey of the sugar industry but, in the end, could not travel in the time frame the NRC was hoping for. It was important that he get there during the early winter in order to make his observations during the sugar harvesting and processing season. When, in mid-November 1945, he learned that LSU would not permit him to travel until June, Keller put his NRC contact, C. D. Shiah 夏勤鐸, in touch with Ernest A. Rose of E. A. Rose, Inc., a Louisiana sugar company, with whom the NRC was able to make speedy arrangements for travel in the winter of 1946.[67]

Rose submitted a comprehensive report on the Taiwan sugar industry, its potential for development, and the cost of redevelopment in February 1946, followed by a set of follow-up reports in July 1946. Rose found that of the forty-two sugar plants in Taiwan, only seventeen were operational, and 35 percent of the railroad system that served the sugar industry had been damaged. The E. A. Rose report projected a cost of $30 million to reconstruct and modernize Taiwan's sugar industry. Costs would go toward the modernization of cultivation practices, including more extensive use of fertilizer, the rehabilitation and modernization of twenty-eight of the processing plants and their transportation infrastructure, and the training of a new managerial workforce to replace the Japanese managers who were departing Taiwan.[68]

By September 1946, both the Taiwan Sugar Corporation and NRC officials were pessimistic about their ability to secure the $30 million that E. A. Rose had projected would be necessary to complete the project and were trying to make do with whatever they could come up with. That August, C. H. Huang 黃振勳 of Taiwan Sugar had written to C. D. Shiah of the NRC that he was "working on the particular project of repairing all damaged factories."[69] Channan Shen 沈鎮南, head of the company, wrote that their human and financial resources were such that they could really only afford to undertake "temporary and urgent jobs" and that they desperately needed the help of E. A. Rose to make permanent repairs. He committed, however, to proceeding with the program that Rose had laid out regardless of whether Taiwan Sugar could secure the funding or not.[70] Already by the spring of 1946, Taiwan Sugar had found ways to purchase fertilizers from the United States, Canada, and Britain, for example, and although delivery of that fertilizer was significantly postponed, the company nonetheless expected a much brighter future in terms of tonnage of sugar cane production than Rose had projected in his reports.[71]

The NRC, meanwhile, continued to explore options that they believed would either help Taiwan Sugar to get a loan from the Export-Import Bank or otherwise secure the company's future. To that end, in the fall of 1946, the NRC engaged in a correspondence with Bitting Incorporated, sharing with Bitting the Rose reports and other relevant data about the sugar industry in Taiwan. That November, Bitting presented the NRC with a proposal to take over the entire management of Taiwan Sugar. There is no evidence to show whether the NRC truly considered acting on this proposal, but it was nonetheless one of the options they had as they tried to figure out how best to get Taiwan's sugar industry up and going.[72]

By January 1947, the NRC had clearly decided not to contract the management of Taiwan Sugar out to Bitting and were asking Rose to help them identify a sugar engineer who could spend time in Taiwan working with Taiwan Sugar. It is unclear whether Rose had any hand in the eventual selection of Robert Ely, a sugar engineer with seventeen years of experience, but the April 1947 contract that governed his activities was made with the General American Transportation Corporation, a company with which the NRC already had a contract to do cement work in Taiwan. Ely traveled to Taiwan in June 1947, stayed about a month, and then returned to the United States to gather a group of five additional men to form an "American Advisory Group" that went to Taiwan in August 1947 to do further surveying and serve as technical consultants for the Taiwan Sugar Corporation. Ely submitted reports in October 1947 and January 1948. These reports both reinforced and supplemented the Rose reports. It seems likely that the primary motivation for hiring Ely and the American Advisory Group, other than that they might provide useful technical assistance while on the ground in Taiwan, was to come up with more evidence from a second source that would corroborate the Rose report, which the NRC was using as a foundation for its loan request to the Export-Import Bank.[73] C. D. Shiah referred to Ely and the

American Advisory Group and summarized their report findings in a letter to Clarence E. Gauss, former U.S. ambassador to China and a member of the board of directors of the Export-Import Bank. He also wrote, "We have engaged a reputable engineering firm, namely the General American Transportation Corporation, as our over-all consultant in regard to the engineering and purchasing for the Sugar Corporation. . . . We are fully aware of our shortcomings in regard to managerial as well as technical personnel, and we are only too glad to accept outside friendly advice."[74] Whether Ely and the American Advisory Group were strictly necessary to help Taiwan Sugar get back on its feet, Shiah and the NRC clearly believed that their report, along with their connection to General American Transportation Corporation, could help with regard to the loan and made a point of mentioning it to Gauss.

By December 1947, the NRC was reporting in a revised loan proposal to the Export-Import Bank that Taiwan Sugar had gotten a $5 million loan from the Bank of China, $1.7 million of which was spent for railway equipment and the remainder for machinery and equipment for sugar factories. The company had also procured 88,000 tons of fertilizer. "The Taiwan Sugar Corporation has thus made considerable progress and its accomplishments are quite outstanding. In the original report, E. A. Rose Inc. estimated that the production of sugar in the year 1947–48 would be around 93,940 tons. However, we are positive we are going to do far better than that. During the coming grinding season, which starts December 1st, we are going to produce 200,000 tons of sugar."[75] Moreover, Taiwan Sugar was having these successes while not even entirely following the advice of E. A. Rose, who had suggested that they dramatically reduce the number of sugar plants on the island to twenty-eight rather than forty-two. Instead, the company was operating thirty-six plants, though undoubtedly none at the level of efficiency that Rose had hoped they would reach under his company's guidance.

In the case of Taiwan Sugar, we see that the NRC contracted for services with more than one U.S. company, both of whom provided analyses of the situation and plans for how to improve it. Existing documents do not clarify exactly how long Ely remained in Taiwan or the extent to which either company provided services beyond the initial planning and advice. Given that the NRC was hampered by its inability to secure U.S. loans for the proposed activities, the prospects for full execution of the

Rose plan under the guidance of E. A. Rose were bleak. Nevertheless, Taiwan Sugar did act on the advice it had received to the extent it could afford to with the resources it was able to amass. It did so in particular with regard to the rehabilitation of sugar railroad transportation, the use of fertilizer to increase yield in sugar cane fields, and the implementation of a training program (which will be discussed later in the chapter).

TRAINING

At least some of the contractors hired by the NRC expressed concern about China's industrial management issues and the impact that a lack of suitable personnel could have on China's ability to carry through any plans that the contractors might help them devise. Glen M. Ruby, of Hoover, Curtice and Ruby, a firm that was engaged by the NRC to investigate oil and gas development opportunities, wrote in a letter to Locke that he was somewhat skeptical of the Chinese capacity to extract and utilize oil should they find it. "If the Chinese had the necessary managerial ability, it would be another matter," he observed, "but in my short visit it was very evident to me that they are sadly lacking in this very important phase of all business."[76] Ruby stated that he had heard similar comments from contractors working on oil refining and hydroelectric power. So strong were Ruby's feelings on this matter that he went on to observe that "we would definitely not be interested in doing any work in China in which we did not have complete managerial powers during the life of the contract. I suggested [to the NRC] that this contract be for a minimum of five and a maximum of ten years during which time we would train Chinese nationals in all phases of the work so that at the expiration of our contract they would be in a position to continue the production without any break."[77] In other words, training of managerial and other personnel would have to be one of the dimensions of any rehabilitation contract with Ruby's firm, but in the meantime, the project would be entirely under American management. Ruby made no mention of the fact that such a contract would almost certainly be financially lucrative to his company.

Ruby was not the only consultant to propose putting Chinese operations under American management on a temporary basis. Pierce Management, the coal mining company mentioned earlier in the chapter, outlined a program for the rehabilitation and modernization of existing mines that called for "actual management, where necessary, of mines."[78] Both Pierce and Ruby saw such American involvement as necessary to safeguard American investments in China. As Pierce put it, "As a protection of American loans, American engineers and particularly American management should be a requisite to provide the necessary driving force to accomplish the planned objectives."[79] Pierce, who appears to have had an exceptionally positive relationship with Chinese government officials, nonetheless repeated some of the standard tropes about Chinese management in his writings to Locke. Officials, he noted, were overburdened, but an even more important problem was that they were "selected for their academic achievements rather than practical business experience." Worse still, "the running of the plants is entrusted to young engineers of little experience. There is not sufficient desire to enter the mines and get dirty, and therefore, in the final analysis policy decisions are taken on the reports of foremen."[80] In other words, in Pierce's view, Chinese industry lacked managers who had developed an understanding of their industries through hands-on experience.

Pierce blamed what he called the "class system" for China's management problems. Believing strongly in the technical capacity of Chinese laborers who, he said, were "well disciplined and very skillful in the use of their hands in making any kind of mechanical device," Pierce argued that to address the management problem, industrial leaders "must come from the masses, which, of course, means the elimination or at least the suppression of the class system. This must be brought about through the elimination of the system of employing friends and classmates, and a reorientation of the mind with respect to the dignity of work." To achieve that end, he suggested to Locke that China should implement "an educational program that is compulsory through high school, and practical courses similar to the International Correspondence School to upgrade the foremen class to provide the corporals and sergeants to execute plans. Also, there must be 'in-plant' training, utilizing the shops, mines and factories to upgrade the better workmen to foremen." He went on to suggest "rigid laws against graft and squeeze, making this law so drastic that it will prevent exploitation for private gain, this in order to encourage the masses to greater effort and create a spirit of National pride, which is essential to the success of a National Industrial Program."[81] Creating pathways for upward mobility within industry and supported by educational programs, inspiring worker commitment and confidence by eliminating graft, squeeze, and old boys' networks, and placing only highly qualified technicians with actual hands-on experience in management positions seemed to Pierce to be the best approach to transforming China's industrial landscape. Without good management, according to Pierce and Ruby, even well-planned industrial development was unlikely to succeed. Such change could not be implemented on a wide scale and probably only rarely on a small scale under the circumstances in which China was trying to revitalize industry immediately after the war. Other consultants proposed more doable training programs for their industries, but even those were sometimes hard to pull off. Training programs that would have taken place in China by and large appear not to have taken place.

The case of Taiwan Sugar illustrates both the ambitions that consultants sometimes had for training of managerial workers and the reality of what occurred. Developing a strategy to increase the number of skilled sugar technicians in Taiwan was an important dimension of the E. A. Rose proposal. As Rose pointed out, there had been roughly six thousand Japanese administrative and supervisory personnel working at the forty-two different plants that now comprised the Taiwan Sugar Corporation, and "native Taiwanians were not educated, trained nor permitted to hold other than very minor positions in the organizations, so that with the departure of the Japanese there is an almost total lack of competent personnel to carry on the multifarious activities essential to the successful conduct of this business." Moreover, the plants had "operated on the basis of an abundant supply of good and cheap labor. No thought seems to have been given to the elimination of unnecessary labor through the installation of more modern and efficient equipment."[82] So, although many Taiwanese worked in the most basic jobs in the sugar industry, few, if any, had been trained for managerial work, and the lower-level workers had been trained in an environment that did not prioritize efficient use of human resources. This meant that there was an almost complete lack of trained technicians to take over the managerial work after the departure of the Japanese.

For that reason, Rose proposed an immediate program whereby twenty graduates of mechanical or chemical engineering departments with leadership skills be sent to Louisiana in the fall of 1946 for a one-year training program to learn all dimensions of the sugar industry.[83] This program would be followed by a long-range initiative wherein two to six students would be sent abroad for training every year for two years of graduate study and an additional year of on-the-job training. Rose's idea was that these students would serve "as an insurance measure to insure progress in the industry," because they would "serve as a source of information on developments outside Taiwan and this will be invaluable in keeping the industry continually moving forward to better and more profitable operations." In addition, Rose argued that "a sugar school should be established in Taiwan as soon as conditions permit. The equipping and staffing of such a school is quite a difficult problem and quite costly. Such a school would serve to train graduates of Chinese or other universities in the practical operation of sugar plantations and factories. Chemical and mechanical engineering graduates with one year of graduate training in such a school could enter the industry in minor positions where they would serve while acquiring the experience and background necessary to qualify for more responsible positions."[84] Rose's proposed sugar school would top off the education of college graduates in appropriate fields and train them up for the specific kinds of work that the sugar industry would require of them.

The leadership of the Taiwan Sugar Corporation agreed that it would be a good idea to send men to the United States for training as soon as possible, though they were initially concerned about whether they could afford to send the full complement of twenty because of their general lack of personnel. By November 1, 1946, however, Channan Shen had let the NRC-NY know that he had a group of twenty engineers ready to go, and throughout the winter of 1946–1947, C. D. Shiah was negotiating with Arthur G. Keller at LSU on the terms of such a program.[85] Keller proposed a one-year program consisting of study at LSU and on-the-job training at the high rate of $38,000 for the group, not including transportation costs.[86] Although the Taiwan Sugar Corporation and the NRC-NY continued to push for this program through the spring of 1947, ultimately the NRC in Nanjing nixed the program because it cost too much.[87] Shiah presented this turn of events to Keller as a postponement due to unforeseen circumstances.[88]

Keller had already trained some sugar technicians as part of the ITA training program in 1945 and 1946. During this period, Huang Chan-hsun 黃振勛, Chen Che-pin 陳濟平, An Chun-tsan 安君贊, and Chang

Lie-tien 張力田 all trained at LSU's Audubon Sugar Refinery and a number of other sugar plants and refineries. There is little doubt that these trainees were already being groomed with the expectation that they would be working with the Taiwan Sugar Corporation. Trainee Chang Lie-tien wrote in one of his training reports about ways in which he could see applying his learning about modern sugar processing methods to the improvement of "efficiency of operation and the production" of the sugar industry in Taiwan. Chang, however, never applied his learning in the way he imagined in his training reports. Although his official training program continued at least until the spring of 1947, Chang continued on at LSU, graduating with a PhD in 1952, and then working in a U.S. firm until 1958, at which point he returned to China and took up a government post in Beijing that used his training in food chemistry.[89] By September 1946, Huang, on the other hand, was already working in Taiwan Sugar's engineering office and was, himself, eagerly pushing for the realization of the training program that E. A. Rose had proposed.[90] Chen Che-pin, who was born in Fujian and had a 1943 college degree from National Central University in agricultural chemistry, also seemed like a natural fit for sugar work in Taiwan. However, at LSU, Chen found himself stuck in the laboratory conducting control experiments for Keller rather than spending time in factories observing their operations. From his perspective, this was not the training he was hoping for as he did not even have time to attend classes because of his laboratory duties, and he was eager to get out into the field. Keller, on the other hand, would only release him to do so once a replacement could be found to conduct the laboratory experiments.[91] Although Keller was clearly generous with his time and LSU's resources in agreeing to take on so many Chinese trainees, both Chen's story and the high price tag that Keller later put on the proposed Taiwan Sugar training program suggest that he did seek, and in some cases found, ways to benefit from their presence.

The "postponed" Taiwan Sugar training program never materialized as planned. However, by late 1947, the NRC had managed to send at least some of the twenty selected trainees to LSU for a yearlong training program anyway. Rather than saying they were part of the new training program, the NRC-NY represented the trainees as delayed members of the original FEA/ITA program that Keller's earlier trainees had come on. As William L. C. Hui of the NRC-NY wrote to Taiwan Sugar trainee Yuen

Yin-kee 袁炎基 on October 31, 1947, "You had better inform Dr. Keller that you are one of the original FEA or MEA technicians who was suppose [*sic*] to have come here two years ago but was detained until recently; now you were sent here by the Ministry of Economic Affairs, for a year's training. Since Dr. Keller had requested a large amount of expenses for the training of twenty engineers from the Taiwan Sugar Company sometime ago and the negotiations were suspended without any conclusion, I did not mention the fact that you were one of the twenty engineers in my original letter to him. You therefore, better not mention anything about this to him." Keller was not immediately responsive to all of Hui's correspondence about additional trainees, but in the end, he did accept at least one more, Chen Tung-yuan 陳東元, who arrived at LSU toward the end of 1947.[92]

The case of the Taiwan Sugar training program illustrates just how much the context of transnational technical assistance had changed in the postwar era. During the war, U.S. corporations, academics, and government agencies had agreed to take in trainees at the request of the U.S. government and, to a more limited extent, groups like the China Institute, which facilitated citizen-to-citizen engagement between the two allied nations. Members of the private sector had agreed to participate out of a sense of patriotic duty, although they sometimes also benefited from the skilled labor that the trainees provided and an even greater motivator was almost certainly the expectation of postwar economic gain. With the phasing out of U.S. government involvement and the introduction of contracts negotiated directly between the Nationalist government and U.S. industries and academic experts, the nature of the relationships changed. Now U.S. actors made it very clear that they expected to gain financially from the relationship, as Keller's proposed training program shows. The Chinese government did not always have the resources to pay the fees that their U.S. partners hoped to get, however, and sometimes had to prevaricate in order to get the result they needed. Scarce resources also meant that the programs suggested by U.S. consultants could not be fully implemented, and the Chinese government was forced to do less and train fewer people. So even in the case of Taiwan, where the Nationalist government had considerably more control over the broader economic and political context for postwar redevelopment than they had in other parts of China, redevelopment projects could not be executed to the utmost.

Continuing Government-to-Government Missions

Although a great deal of the technical interaction between China and the United States switched, on the U.S. side, to the private sector with the end of the war, bilateral technical relations did continue between government entities in some areas, most notably agriculture. In the fall of 1945, China's MOAF requested that the United States send an agricultural mission. In early 1946, the U.S. Department of Agriculture agreed to send an eight-person group to China to work with Chinese agricultural experts to devise a plan for agricultural development in China. The mission would include experts on tung oil, tea, soybeans, silk, carpet wool, and fisheries who would work out recommendations for continued collaboration following the end of the mission.[93] The decision to participate was undoubtedly influenced by President Truman's stated enthusiasm for the development of agriculture in China, which he believed would help that nation to find political and economic stability. The U.S. Office of Foreign Agricultural Relations did not want the mission to be a waste of resources, however, and was particularly concerned, as U.S. advisers to China tended to be, that China have a "strong central agency" that would be capable of implementing the mission's recommendations.[94]

The nine-person mission, which was led by C. B. Hutchison, dean of the College of Agriculture at University of California at Berkeley, and included agriculture experts from both the U.S. Department of Agriculture and several U.S. universities, arrived in China on June 27, 1946. For several weeks in Shanghai and Nanjing, the U.S. group met with the thirteen-person Chinese team headed by P. W. Tsou. After spending several weeks in conferences in Shanghai and Nanjing, they split into groups to survey fifteen provinces, including Manchuria and Taiwan. The mission's principal recommendations included integration of agricultural education, research and extension under the MOAF and Ministry of Education, creation of a new banking and credit structure to provide financial support for agriculture, greater emphasis on regulation of agriculture, greater emphasis on fertilizer, irrigation, and plant and livestock improvement, and expansion of agricultural exports.[95]

Owen L. Dawson, U.S. agricultural attaché in China, writing to Ambassador Stuart in China just after the conclusion of the mission in late

November 1946, indicated that he saw this collaboration as having the potential to set a model for future U.S.-China relations. "This is the first mission of its kind bearing upon possible economic collaboration between China and the United States," he wrote, "and its success, if measured by significant results in following out its recommendations, will be a most hopeful sign looking toward any future plans or arrangements for assisting the Chinese Government in educational, economic, and allied fields."[96] Dawson was viewing the experience from within China and may not have had a very clear sense of the extent to which the U.S. government was, in other ways, pulling away from its wartime pattern of support for technical missions in China.

There was considerable enthusiasm for the mission on the part of leaders of China's agricultural institutions, especially T. H. Shen, who had advocated for and implemented a variety of changes with regard to agricultural research and extension during the war and had actually been working with Dawson for several years to plant the seeds for the mission with their respective higher-ups.[97] By late November, Shen had produced and provided to Dawson a plan for immediate steps that the MOAF should take, particularly with regard to extension work.[98] Shen and the other mission members went on to produce a substantial *Report on the China–United States Agricultural Mission*, which was intended as a guide to future action in the field. Like Dawson, Shen regarded the outcome of the mission very positively, later writing in his autobiography that the "experience with the Mission increased my belief in the Chinese-American joint organization."[99]

As we have seen, technical cooperation organized and carried out by the two governments did continue in the postwar period, but it ceased to be the dominant model. Increasingly, the Chinese government had to look outside of government circles for the technical assistance it required. The new postwar model for technical cooperation was based on economic transactions, albeit ones that were sometimes brought about because of human relationships rather than common political goals. Locke ceased to have China's economic development as part of his portfolio in the White House, although he continued to serve as an economic adviser to the Nationalist government and kept Truman abreast of developments in China as he understood them. The general attitude toward China in U.S. diplomatic circles was increasingly one of frustration with the

Nationalists for failing to implement good government (as understood by those diplomats). Chinese diplomats practically had to beg for loans to support their rehabilitation efforts. As Wang Shou-chin of the CSC wrote in a February 1947 memorandum asking for disbursement of funds that had been earmarked for China by the Export-Import Bank and expedition of other matters such as the distribution to China of Pacific surplus items and of Japanese industrial machinery as war reparations, "The present economic situation in China is not beyond recovery if effective action is taken now showing confidence and constructive steps by the United States."[100] On the whole, however, the U.S. government failed to take such steps with regard to postwar China. As China's internal economic and political/military situation worsened, the Nationalist government was left with quite a large number of good and expensive plans and an increasing number of well-trained technicians but relatively little, at least in most regions, in the way of successful implementation of those plans by those technicians.

Conclusion

The end of the war did not signal an abrupt end to all of the cooperative technical programs that China and the United States had developed over the course of the early 1940s, but shifts in U.S. priorities did lead to changes in how those programs operated and other changes in the strategies the Nationalist government had to employ to get U.S. assistance. The FEA/MEA technical training program had started out as a lend-lease-funded program designed to build a cadre of technicians who could work to improve the efficiency and productive capacity of various inland industries and who would be well prepared to undertake immediate technical challenges in areas such as rehabilitation of ports and railroads at the end of the war. The second group of technicians in the program was only just reaching the United States as the war ended, however, so those two aims were no longer entirely appropriate by the time their training programs actually started. Now trainees were being prepared to rehabilitate or build industries, primarily in formerly occupied areas. The actual training programs that trainees undertook probably did not change much as a result of the shifting context, but one thing that did change was the expectation that trainees had of what they would do after their programs ended. Whereas trainees were selected for the program with the expectation that they would return, for the most part, to their old NRC factories and would have had a very clear sense of just what those units needed, now they did not always know where they would end up or what the specific needs of those sites would be. Nonetheless, the program trained and returned to China hundreds of technicians in a wide variety of technical and industrial management specializations. The trainees not only gained hands-on experience with successful American enterprises in their fields, but they also developed relationships that many of them surely continued to maintain in the coming years. Also worth noting about this program is the fact that the NRC continued into the late 1940s to use its structure as a vehicle for embarking on new, more-targeted training programs for specific industries, such as sugar, as the need for such training became apparent.

With regard to planning for and undertaking the actual development of industry, however, we see a bit less continuity. Even before the AWPM came to an end, the NRC was seeking to establish relationships with individual U.S. industries for the purpose of gaining both general planning advice and specific rehabilitation assistance. The NRC clearly expected that industry-specific plans drafted by reputable American firms and based on both on-the-ground observations and reliable data would help them secure U.S. rehabilitation loans. To that end, they included such plans with their applications to the Export-Import Bank and in some cases even got a second opinion to further bolster their applications.[101] A combination of circumstances on the ground in China and the Export-Import Bank's failure to grant the loans often made it impossible for those plans to be put into effect. However, throughout the period of 1945–1949, the NRC continued to plan for their implementation, and their branch office in New York played an important role in maintaining relationships with both the industrial consultants and friends in the U.S. government. In the meantime, NRC organs in China forged ahead to implement such aspects of those plans as they were able to undertake with the resources they could find. In other words, the plans were not simply solicited as a strategy to gain U.S. funds. To the extent possible, the NRC carried through with them and used them as a foundation for further action.

An important postwar shift was that U.S. technical consultants were no longer being sent to China under the auspices of a U.S. government program. Now the NRC-NY and the CSC in Washington, DC had to identify such consultants for themselves (though they typically sought advice in such matters from contacts they had made through earlier U.S. government programs) and negotiate their own contracts directly with the consultants that stipulated the scope of work to be done and the fees that would be paid for that work. Through this process, the NRC gained agency in a process that had been largely dominated by U.S. interests during the war. Whereas the cultural relations program had at least sent experts in areas requested by the Nationalist government, the AWPM had aligned its activities with U.S. government priorities. Through the new contract system, the Nationalist government could once again bring in consultants for projects that it prioritized and could guide them to work in accordance with those priorities. The consultants may have found some of the Nationalist government's priorities to be problematic, but they were getting paid to serve that government, and they provided plans and advice aimed at helping the Chinese government to realize its goals.

The consultants operated primarily as representatives of their own businesses and always had potential benefits to those businesses in mind as they made their recommendations. As a result, recommendations were sometimes inflated in terms of both what needed to be done and how much it would cost. Other consultants disparaged Chinese technical and management capacity and made clear that they would only undertake to implement their plans if they had full or nearly full control of the projects. Chinese observers were also concerned by China's scarcity of personnel trained in industrial management, but even so, the reluctance of some consulting firms to work with Chinese managers might also have been part of a bid to have a greater say over their projects and to earn more revenue from them.

Epilogue

With the end of the war, most of the direct state-to-state mechanisms through which the Nationalist government had received wartime technical training and advice from the United States ended. Although the Foreign Economic Administration (FEA)/Ministry of Economic Affairs (MEA) training program that had been set up during the war continued and, in fact, grew considerably in the two years immediately following the war, the FEA ceased to be involved and coordination of the program in the United States was taken over by the International Training Administration (ITA), an independent entity that had spun off from the U.S. government. Postwar U.S. aid to China was funneled primarily through the United Nations Relief and Rehabilitation Administration (UNRRA), a UN-managed agency headquartered in Washington, DC and overseen by an international group of administrators that included, but was not limited to, both Chinese and American personnel.[1] In addition, new postwar institutions came into being, such as the Joint Commission on Rural Reconstruction (JCRR), a joint U.S.-Chinese organization aimed at helping China, and later Taiwan, to develop its agricultural capacity. The JCRR was a semi-independent organ that was formed as part of the 1948 China Aid Act and its members were all appointed by either the Republic of China (ROC) or U.S. government, but its activities were explicitly understood to be independent. In all respects, the U.S. government was moving away from the kind of direct involvement in supporting Chinese technical projects that had characterized the war-time interactions between the two nations.

After the war, therefore, China could no longer rely on the kinds of coordinating assistance that it had previously received from the U.S. government. In order to carry out postwar reconstruction plans, the Nationalists were compelled, instead, to start working directly with American businesses. In the waning months of the war and thereafter, the Nationalists set up numerous contracts with American businesses for consulting services and also to purchase materials. This new model of Sino-American technical engagement gave the Nationalists considerably more decision-making autonomy than they had had under the American War Production Mission in China (AWPM)/Chinese War Production Board (CWPB) arrangement. The business contracts were determined entirely on the basis of the Nationalists' own priorities. Of course, the businesses with whom they contracted also had their own priorities and motives that generally involved not just earning whatever was agreed to in the contract but also making longer-term inroads into the Chinese market. In spite of the fact that the Nationalist government acted swiftly in the summer and fall of 1945 to contract with businesses that they hoped would be able to help them with the postwar rebuilding process, only parts of the large-scale economic development plans that the Nationalist government produced during the war came to fruition and only some of those efforts began to bear fruit while the Nationalists were still located in China. Nonetheless, the story of Sino-American wartime technical cooperation can hardly be called a failure. To the contrary, it laid important foundations for later industrial and agricultural development efforts in both China and Taiwan.

It would be easy to dismiss the efforts described in this book as having been wasted and ultimately meaningless. After all, the Nationalist government lost the civil war and retreated to Taiwan and few of their plans for utilizing the technical knowledge and capacity they had built during and immediately after the war were realized. But these efforts did have consequences, if not always the intended ones. In addition to laying both human and infrastructural groundwork that facilitated later technological development in the People's Republic of China (PRC), these efforts also reveal some of the structural origins of what would later become known as the developmental state in Taiwan. In that sense, the efforts described in this book constitute the antecedents of what has been called the Taiwan miracle and, in a much more limited way, the PRC's own economic development.

China

Many of the specific wartime projects the Nationalist government undertook in concert with its American technical advisers did yield mid- to long-term results, albeit mixed ones. The iron and steel industry provides a good example. The plant at Shijingshan that had been the first step in the iron and steel development plan that Arthur G. McKee and Company and the National Resources Commission (NRC) worked out in 1946 remained an important location for iron and steel production well after the PRC was established. This site predated the war and had been taken over and more fully developed by the Japanese during the war, though it clearly needed rehabilitation as of late 1945. By the time the McKee consultants visited the site in April 1946, the NRC was already hard at work rehabilitating the plant, though not, in the view of the McKee consultants, in ways that were well designed or highly functional. By 1947, McKee had contracted with the NRC to provide certain machinery for the Shijingshan plant, and correspondence from 1948 suggests that the machinery had been received and was being installed.[2] Although McKee almost certainly did not see through the complete overhaul of the plant, the company did fulfill at least some of its contractual obligations and contributed to its rehabilitation. Following China's political transition in 1949, the plant continued to operate and the Central Intelligence Agency (CIA) was still tracking its activities at least as late as November 1965, noting in a surveillance report that the plant was still active and had three large blast furnaces as well as rolling mills, coke ovens, and other by-products sections.[3] A 1956 report on China's iron and steel industry also noted that the Chinese Communist Party was at that time building "a large, fully integrated iron and steel mill" at the Daye site for which McKee had produced blueprints for precisely the same sort of mill, though we do not know whether the PRC was utilizing those blueprints. Nonetheless, both Daye and the Chongqing region were still, in the 1950s and beyond, important production points. In other words, although the work done by the NRC to build mills in the inland region, and later by Arthur G. McKee and Company as requested and paid for by the NRC, to rehabilitate and plan for the construction of mills in recovered regions ultimately did little to support Nationalist economic development plans, those plants were important to economic development under the PRC in the 1950s, '60s, and beyond.[4] It is difficult to assess just how meaningful U.S. assistance was in the wartime improvement or postwar rehabilitation of these plants, but it is still worth noting that the plants themselves along with many of the technicians and laborers who worked in them did play an important role in China's post-1949 industrial development.

Other kinds of wartime scientific and technical institutions also survived the war and played roles in the postwar development of those fields. For example, the China Institute of Geography, established in 1940 in Beibei, Sichuan, was absorbed into the Chinese Academy of Science as its Institute of Geography in 1950.[5] The northwest branch office of the China Geological Survey served as the model for later branch offices in Taiwan and Manchuria.[6] In other words, some of the structures and the work models that developed under the Nationalists during the war continued to exist and to be influential in the post-1949 era.

It appears, on the other hand, that few of the long-term projects developed by the cultural relations experts yielded large-scale lasting results. Despite Republic of China (ROC) government enthusiasm for Walter C. Lowdermilk's plan to mitigate soil erosion in western China, the U.S. government failed to follow up on it, leaving Lowdermilk with considerable bitterness about the efficacy of his work in China. From his perspective, had he been provided with the resources to pursue his experimental projects to the end, his team's work might have made a real difference in both reducing soil erosion and developing the agricultural potential of the Gansu-Shaanxi-Qinghai region. But such resources were not forthcoming.[7] Dykstra's potato program, on the other hand, may have yielded at least some lasting results. A 2009 study of "Potato Production and Breeding in China" points to the introduction of foreign cultivars (probably by Dykstra) in the 1940s as an important effort, though it also notes that a 1983 study suggested that Chinese potatoes all stemmed from a fairly narrow genetic background. Nonetheless, it seems that at least some of the cultivars Dykstra introduced may have contributed to that genetic material.[8]

Although the outcomes of the labor of Dykstra, Lowdermilk, and other cultural relations experts may have had a much more limited impact than they had hoped, such experts did help to train people in their fields who continued to work in China, Taiwan, and elsewhere. Lowdermilk recalled later, for example, that a number of the young technicians he trained went on to do agricultural extension work, scientific research, and forestry work in China, whereas a couple of the others left for Taiwan where they did similar things. However, one of the men who had remained in China, he had heard, had been "liquidated by the Communists because of his friendship for me and for America."[9] Training experiences fostered by the programs mentioned in this book did, therefore, have very negative consequences for trainees in at least some cases.

The FEA/MEA/ITA training programs also unquestionably had a long-term impact both on the individual trainees and on Chinese agriculture and industry. Although it is beyond the scope of this work to investigate how many of the returned trainees remained in the PRC to contribute in the fields in which they were trained, we know that at least some of them did. Upon his return from the United States in 1946, Jiang Shanxiang, the trainee studying phosphorous mentioned in chapter 5, was sent to serve as the head of the China Match Materials Factory (中國火柴原料廠) in Qingdao. True to the plans he outlined in his final training report, in 1950 he built China's first yellow phosphorous electric furnace in Sichuan and its first calcium magnesium phosphate blast furnace in Beijing in 1957. In the 1960s, he worked to build the first large yellow phosphorous plant in Guangxi. He continued to be a leading figure in China's phosphorous industry for the rest of his life.[10] Ma Shiying 馬世英 of the Central Machine Works in Kunming was sent for training in September 1945 in engine lathe making and ended up staying on to do an MA in applied physics and engineering at Harvard. Following his 1948 return to China, he worked at the Central Machine Works in Shanghai. He eventually shifted to the manufacture of airplane engines at Xi'an's Aero Engine Manufacturing Plant, becoming the first director of the China Aviation Society.[11] Li Qingkui 李慶逵, who had been sent for training at the Illinois Agricultural Experimental Station in September 1945 to study soil nutrients and soil fertility, stayed on and completed his PhD at the University of Illinois in 1948. He subsequently returned to China and by 1955 had been elected to the Chinese Academy of Science. He continued to perform soil and fertilizer research in China into the 1980s.[12] Chang Lie-tien, one of the Taiwan Sugar trainees at Louisiana State University (LSU), remained in the United States until 1958, at which point he returned to the PRC and worked in food industries and as an educator. Weng Xinyuan (翁心源), Weng Wenhao's son and one of the thirty-one, remained in China after 1949 and continued to work in petroleum geology until his death in 1971 in the Cultural Revolution. At least half of the other thirty-one trainees also chose to remain in China after 1949 and worked in the fields they had been trained in. A number of them were elected to the Chinese Academy of Science in 1955 and some continued to work in the same factories.

Like the trainees, many of the men who had been involved in planning and executing the programs described in this book also remained in or returned to China after 1949. P. W. Tsou, the agronomist connected to the Committee on Wartime Planning, returned to China in 1956 and worked for the Ministry of Agriculture. T. P. Hou, also connected to the Committee on Wartime Planning, remained in China, continued to work in the chemical and fertilizer field for various Chinese government agencies, and was named to the Chinese Academy of Science. Weng Wenhao briefly left China in 1949 but returned in 1951 at the invitation of Mao Zedong and Zhou Enlai. He served in the Chinese People's Political Consultative Conference, of which his former NRC deputy, Qian Changzhao, served as vice-minister.

Nationalist industrial training programs in China also produced skilled technicians who went on to work in related industries in the PRC. For example, most of the technicians recruited to work at the aircraft engine factory in Crow Cave at Dading did not have a background in aerospace engineering, and only some of the aerospace engineers, such as Yaozi Li, whom we encountered in chapter 3, had the opportunity to study abroad. The plant therefore had to conduct its own trainings. Ouyang Changyu, a graduate of the Zhejiang University aerospace engineering program and one of the 286 skilled technicians who worked at Dading, located seventy-five former Dading technicians in 1995. Of the technicians he traced, 68 percent had remained in China, 20 percent were university professors, and 40 percent were senior engineers in a variety of industrial sectors. Ouyang himself, along with several other Dading engineers, had spent his career researching and building missiles for the Chinese army.

Only 21 percent of the people he had tracked down in 1995 were Zhejiang University graduates, and the remaining 79 percent had received their training at the factory.[13]

One of the aerospace engineers who was sent abroad from Dading, Wu Daguan (mentioned in chapter 3), decided to go to Beijing upon his return in 1946 and teach at Peking University. After 1949, Wu was instrumental in recruiting some of his former Dading colleagues, including Wang Shizhuo 王士倬, the head of the plant, to work for the new PRC government, first repairing and then later manufacturing Soviet aircraft that the Chinese army was using to fight the Korean War. Wu continued to manufacture aircraft for the PRC military for the rest of his life.[14] Although far from comprehensive, these various examples show that many trainees, Nationalist technocrats, and independent scientists and technicians who had been involved in the scientific and technical development programs described in this book returned to or remained in China after 1949 and contributed to the development of Chinese industries, agriculture, and academia.

As noted above, the American connections that former trainees and students established through the wartime programs in which they participated led in some cases to criticism or worse during the Maoist period. Even so, some former trainees rekindled those connections as soon as they could in the early 1980s. In 1983, Wesley Buchele, a professor of agricultural engineering at Iowa State University who had been a member of the team that had taught in the International Harvester agricultural engineering program in 1945 and 1946, was invited to China to teach a course at the Beijing Institute of Agricultural Mechanization (北京市农业机械化研究所, BIAM). His host and the head of the institute was Li Hanru, himself an Iowa State graduate from the late 1940s who had remained in China after 1949. While in China, Buchele lectured at numerous other major agricultural research institutes and some factories and was interested to discover that the institutes as well as some of the factories were being run by former International Harvester scholars. As Buchele wrote, "Our graduate students had moved to the top of the Research Institutes, University Administrations, and factories and were now the leader in research, academic institutes, universities and factories."[15] These connections not only helped make Buchele's 1983 visit happen, but they also led to his continued interest in and support of BIAM. In addition to

returning to teach there in 1994, Buchele supplied BIAM with back issues of agricultural engineering journals that, according to his students,
had all been lost during the Great Leap Forward. The International
Harvester program did not yield precisely the results that the company
was anticipating. After all, the mainland Chinese market for American-
made agricultural implements closed off with the establishment of the
PRC. However, in spite of and probably also because of their American
training, many of the students from that program had risen over time to
become leaders of China's agricultural engineering community.

As we can see from these few examples, the contributions to later Chinese development made by wartime industries and technicians trained
in wartime and postwar Sino-American programs were considerable, and
the connections they made during their training periods continued to
bear fruit for China many years later. However, we can also find still other
kinds of echoes of wartime technical development efforts in contemporary
China. For example, the Chongqing region had established itself during
the war as a safe zone in which critical industries could operate, often in
underground facilities, even in spite of Japanese air raids. In the 1960s,
the PRC government, concerned that its critical industries were too vulnerable in coastal locations, oversaw a second wave of industrial relocation to Chongqing, moving a number of military factories and even locating a major nuclear facility (816 Nuclear Military Plant) in a massive
new cave that was dug starting in 1966.[16] Many of the wartime factories
in Chongqing continued to function for decades in the postwar period,
though in recent years, at least some of them have been converted to art
centers and cultural museums.[17]

Some of the specific projects undertaken during the war have also
come back to life in recent years. For example, there has been a resurgent
interest in cracking tung oil for use as a biofuel over the past decade in
China. This fuel development strategy, though not actively encouraged
by the AWPM, which favored power alcohol, was actively pursued by the
NRC and was a subject that AWPM technical advisers had weighed in
on both at NRC plants and in the context of the CWPB's Liquid Fuel
Advisory Committee. Over the past fifteen years or so, as part of the
PRC's broader strategy of reducing petroleum dependence and developing alternative fuels for both economic and climate reasons, China's major energy companies have invested in development of alternative fuels

such as ethanol and biodiesel (both used in wartime China). In this context, Chinese energy companies have been exploring the viability of using tung oil for this purpose, leading to an explosion of scientific research on the product. Numerous patents related to the manufacture of biodiesel from tung oil have been filed in the PRC, and at least one energy company in Guizhou contracted with farmers to expand the acreage planted in tung trees.[18]

One example of continuity of wartime technical training efforts all the way to the present day can be found in the continued existence of Bailie Schools in China. The Shandan Bailie School survived the war, though it relocated to Lanzhou and became the Bailie Oil School in 1952. In 1983, a Bailie University was established in Beijing, and in 1987, the Gansu provincial government, working with overseas friends with connections to Rewi Alley and Ida Pruitt, opened a new Bailie School in Shandan that continues to train students in practical skills. According to a 2015 *China Daily* article, the new Shandan school had already trained ten thousand students in its first eighteen years, a number of students that would have been unimaginable to the founders of the early Chinese Bailie schools. Of the first Shandan Bailie School, the new school's headmaster observed that its "students transformed Shandan's productivity level from medieval to industrial age in a few years in the late 1940s" and that Bailie "students also became backbone technicians for the oil fields that developed in the 1950s and 1960s in Gansu, Heilongjiang, Shandong and Liaoning provinces."[19] The Bailie schools thus appear to have left a strong legacy of technical education both in terms of the contributions their early graduates made to development in the region and an educational approach that was consistent with the aims and interests of the Chinese Communist state.

In the Chinese Communist state's approach to science and technology, at least in the 1950s, we also see some resonances with the Nationalist approach of the wartime period. A guiding principle of the PRC's 1956 twelve-year science plan was to "let tasks lead disciplines," or to orient scientific and technical development toward practical application. As Zuoyue Wang has argued, that plan, which served as a blueprint for industrialization, resulted from compromises between the party state and the scientific and technical elite.[20] That scientific and technical elite included some of the people we have studied in this book. Whereas during the war there were still debates, as we saw in chapters 1 and 2, over the extent to which science and technology should be made to serve the interests of state and nation, by the 1950s, this way of thinking had come to dominate China's scientific and technical community.[21] As Wang has pointed out, "the Communist government inherited not only a research infrastructure but also a scientific community dominated by an elite who were Western educated but who chose to stay on the mainland and who were already sympathetic to leftist ideas such as the planning of science."[22] Although most of the activities described in this volume involved technicians rather than the elite scientists Wang was writing about, even so, those activities almost certainly helped to guide these scientists to this way of thinking and led them, in 1956, to call on the government to develop a national science plan.

In each of these examples, we can see ways in which the wartime technical development activities undertaken by the Nationalists and their partners in China were foundational for the development of both human resources and attitudes toward science and technology in post-1949 China. We can also see in at least some of them ways in which transnational relationships that were built during the war were revived in the post-Mao period and formed the basis for further growth and development. We see remnants of that legacy even today, in spite of the fact that the technical relationship between China and the United States is increasingly fraught and the amity and cooperation of the 1940s, which admittedly was undergirded on the U.S. side by currents of opportunism and self-certainty, has been overlooked and forgotten.

Taiwan

Just as the wartime technical relationship between China and the United States proved significant to China's later development, so too was it important for Taiwan. Individual participants in the training and technical development activities discussed in this book migrated with the Nationalist government to Taiwan and put their training and skills to work in that new context. In addition, the patterns of Sino-U.S. technical engagement that were established during the war formed the basis for continued engagement and had a significant impact on both the construction and the efficacy of Taiwan's developmental state. To a certain degree, then, one could argue that the so-called Taiwan miracle resulted from the technical relationships that the United States and China developed during the war.

As we saw in chapter 5, the rehabilitation of both agriculture and industry in Taiwan was a top priority for the Nationalist government in the immediate postwar period. To that end, the government utilized its large-scale U.S. training program to train at least some personnel in the specific industries that were important in Taiwan and also engaged with U.S. industrial consultants to assist in developing plans for the rebuilding of those industries. The case of sugar, as we have seen, serves as an excellent illustration of how the NRC sought to use its U.S. connections to rebuild an important industry in Taiwan. After the Nationalist retreat to Taiwan in 1949, sugar turned out to be even more important than originally anticipated by the NRC as it was one of Taiwan's most abundant products. By the mid- to late 1950s, Taiwan was exporting sugar to other places in Asia including Malaysia, Japan, and Korea as well as to other countries such as Germany and South Africa, so that by the latter part of the decade, the foreign exchange that Taiwan acquired from its sugar sales represented 60 percent to 70 percent of Taiwan's total exports and was, according to sugar researcher Chwei-lin Fan, "the major financial resource of that government."[23]

Fan was an economic researcher with the Taiwan Sugar Corporation and Taiwan Sugar Experiment Corporation who had spent the 1950s working directly with sugar farmers all across Taiwan to "discuss farm management problems" and collect survey data.[24] In 1959, Fan was brought to the United States by a grant from the Council on Economic and Cultural Affairs (later renamed the Agricultural Development Council) to study the economics of sugar production in Taiwan. During his two years in the United States, Fan spent time at LSU and the University of Minnesota (both institutions had participated in wartime and immediate postwar training programs for Chinese sugar and agriculture technicians, respectively), and completed his MS in agricultural economics at Montana State University in Bozeman. He later went on to complete a PhD in agricultural economics, still focusing on sugar, at the University of Hawai'i in 1967.

Fan's MS research investigated a set of price stabilization policies that the Nationalist government had implemented for sugar in the 1950s with the intent of encouraging farmers to plant sugar cane, even at moments when the global price for sugar was depressed, by setting minimum prices for sugar. Such an approach, though not fully spelled out, had been proposed by E. A. Rose, Inc. in its 1946 consulting report on the Taiwan sugar industry.[25] In other words, the consulting reports for which the NRC had paid in the immediate postwar period continued, at least to some degree, to serve as guides for action in Taiwan for some time after 1949. In addition, the networks that the NRC had developed during and immediately after the war for the acquisition of technical machinery also continued to function after 1949. For example, the Taiwan Sugar Corporation continued to purchase parts for sugar refineries in the United States through the NRC-NY in the early 1950s, and as we have already seen in the case of Chwei-lin Fan, the academic and training pipelines the NRC had developed in the 1940s continued to function more than a decade later.[26]

Although many of the trainees the NRC had sent to the United States remained in China after 1949, others did go to Taiwan along with numerous NRC technicians who had not gone abroad for training. Perhaps the most famous trainee to head to Taiwan was Sun Yunsuan 孫運璿, one of the thirty-one, who served as ROC minister of economic affairs from 1966 to 1978 and then as premier until 1984, overseeing a critical period in Taiwan's scientific and technical development. In addition, many of the NRC technicians who had worked to revitalize Taiwan's sugar industry between 1945 and 1949 remained in Taiwan after 1949. Of the technicians working at the Taiwan Sugar Corporation's Xihu plant in 1950, 23.5 percent were from mainland provinces, for example.[27] According to a 1951 report of the Taiwan Power Company, about 34 percent of its employees were from the Chinese mainland and 17 percent of its employees had a college education.[28] Huang Chan-hsun, An Chun-tsan, and Yuen Yin-kee, all of whom had been sent to LSU as sugar trainees, went to Taiwan where they remained active in the sugar industry. Numerous employees of the China Petroleum Corporation retreated to Taiwan from the Yumen oil fields in Yunnan as well as other NRC enterprises that had worked with other types of oils elsewhere in inland China.[29] Some among this substantial number of technicians had surely participated in the NRC training programs either during or immediately following the war. To foster development in Taiwan, however, new technical training opportunities in the United States were created after 1949. For example, the China Area Aid Act of 1950, which was passed just a few weeks before the outbreak of the Korean War and administered by the United States Mutual Security Agency, included funding designated for technical assistance that brought technicians from Taiwan to the United States for training in the early 1950s.[30] Similarly, the pipeline for students from Taiwan to study abroad in the United States expanded after 1950, though the first big wave of such students did not start arriving until 1953 because male students were typically required to complete military service before studying abroad.

The relationship between the Nationalist government and its U.S. consultants, however, did undergo a fairly significant transformation as the Nationalists retreated to Taiwan. In 1948, following the enactment of the China Aid Act, the U.S. Economic Cooperation Administration (ECA) took over responsibility for paying out U.S. aid to China. As part of its strategy for managing aid and determining which projects to fund, the ECA opted to piggyback on the existing Chinese government contract with J. G. White, a firm it described as "a reputable engineering concern," tasking it with guiding reconstruction efforts in Taiwan, which was, at that point, the only part of China where such reconstruction efforts were viable. The NRC had originally contracted with J. G. White on June 22, 1945, to assist with the development of power plants in both Manchuria and Taiwan but had temporarily discontinued the relationship, probably owing to lack of Nationalist government funds for the project, in the summer of 1947. By the end of that year, however, J. G. White was back at work, consulting with the NRC on a proposed telephone plant.[31] The fact that J. G. White was working with the Nationalist government on Taiwan projects at the time of the passing of the China Aid Act almost certainly influenced the ECA's determination that it was the right company to work through, at least for Taiwan projects, though it was not the only U.S. company in that position. Therefore, it seems likely that there were additional reasons for the ECA's choice.

Whereas the early contractual relationship with J. G. White, like Nationalist contracts with other companies, had been directly between the Nationalist government and the company and was supported by either the Export-Import Bank or Chinese government funds, by 1948 the ECA had flipped the relationship. In essence, it took over management of the contract, making it clear that it would rather bypass the Nationalist government to work directly with the U.S. contractor, who, with input from the Nationalists, would undertake or oversee the reconstruction work that the ECA was financing.[32] So the U.S. government was now directly paying a U.S. firm to do work in China, and the involvement of the Nationalist government in the process was advisory rather than managerial. This new system did not sit especially well with other U.S. firms with whom the Nationalists had been doing business but who could not get the ear of the ECA. As of the spring of 1949, James Pierce of Pierce Management, a firm with which the Nationalist government had been working since 1945, was complaining to L. F. Chen of the NRC-NY that the ECA had retained J. G. White "for the specific purposes of passing on pre-projects from the engineering and economic angles, in addition to the amount of money required for any specific project."[33] So at that point, J. G. White was already beginning to operate as a sort of clearinghouse for the ECA. From Pierce's perspective, this was yet another impediment in a long string, not the least of which was the political situation in China, to work his company had been planning for quite some time, and it is evident that he found it frustrating. By 1950, however, Pierce's objections were moot. The Nationalist government had fled to Taiwan and mainland projects of the sort that Pierce had planned were no longer under consideration.

Toward the end of 1948, a group of five engineers from J. G. White went to Taiwan to survey the island's industrial capacity and remained there into the 1950s to help the Nationalist government rehabilitate factories and develop a wide array of industries such as textiles, plastics, and timber.[34] With the encouragement of the ECA, J. G. White essentially became Taiwan's general contractor. It provided engineering and management services to assess the viability of projects, repurpose machinery, and advise on management and efficiency, and it also provided technical training services for some projects. After the Nationalist retreat to Taiwan, given the existing relationship between J. G. White and the ECA and given also that the ECA contended that using a single company to assess projects reduced costs, J. G. White was well poised to serve in the principal contracting role for ECA-funded projects in Taiwan.[35] Consequently, the relationship between the Nationalist government and J. G. White blossomed after 1949. The company oversaw most of Taiwan's major development projects in the early years after the Nationalist government's retreat and the Chinese side of that relationship was managed by the Council for U.S. Aid (CUSA).

The role played by J. G. White was, in fact, not dissimilar to that of the AWPM during the war. Like the AWPM, J. G. White's focus was on the development of critical industries. The issues that Taiwan faced in the early 1950s were in many respects similar to those the Nationalist government had faced in wartime inland China: economic stabilization, support for a population that had recently seen a major increase owing to the influx of refugees, import substitution, reconstruction, and self-defense. In some ways, for any CUSA staff who had previously been with the NRC or the CWPB, it must have looked like déjà vu all over again.

A major difference, however, was the institutional mechanism that the United States was employing to mete out technical assistance. This time, the U.S. government was paying a private firm to manage the process rather than fielding its own operations that commandeered staff from the private sector and sent them to work on behalf of the U.S. government in China. This shift from public to private was an intentional strategy on the part of the U.S. government, one the FEA had been aiming toward since before the end of the war. As we saw in chapter 5, it was also, at least partly, the result of an intentional strategy on the part of the Nationalist government, which, in the spring and summer of 1945, had started contracting with U.S. firms to provide technical advice. The NRC could see the limitations of the AWPM model and, looking ahead to the postwar period, believed that by working with U.S. firms on important reconstruction projects, they would be more likely to secure the aid they felt they needed to undertake those projects. That strategy did not really pay off until after the Nationalist government fled to Taiwan, though it did not pay off in the way the Nationalists had hoped. With the ECA's insistence that just one firm should serve as general contractor, the Nationalists' agency was diminished. Of the multitude of firms with which the NRC had contracted between 1945 and 1949, there was only one with which the ECA was willing to work, and the Nationalist government's control over whom it would contract with and on what projects was reduced. The shift to the general contractor system probably also had an impact on U.S. businesses that were not favored by J. G. White. The frustration that James Pierce expressed in the spring of 1949 had been justified although, of course, none of the projects he had been working on with the Nationalist government could have come to fruition anyway because of their location.

The relationship between the Nationalist government and J. G. White underwent a number of changes over time, but it remained both intact and central to a wide array of Nationalist development efforts, particularly, but not limited to, efforts financed by U.S. aid, into the 1960s.[36] J. G. White's corporate contract with CUSA ended in July 1962, but even so, the company continued to work with the Nationalist government— for example, contracting six J. G. White engineers to work for CUSA's engineering consulting group on railroad reconstruction after 1962.[37]

The example of J. G. White demonstrates how wartime and immediate postwar relationships continued and in turn begot new relationships that would also, as I showed in my first book, play a meaningful role in assisting Taiwan to develop its industrial and agricultural capacity in the 1960s and '70s. The institutional structures within which these relationships blossomed did, however, change. As noted above, the end of the war led to a transition in the institutional framework in both the United States and China that managed U.S. assistance to China so that first UNRRA, then the ECA, and eventually the U.S. Agency for International Development managed the bulk of U.S. aid to China and then Taiwan. On the Chinese side, CUSA was established and placed under the leadership of Weng Wenhao in 1948 to manage U.S. aid that came through the China Aid Act of that year. The new institution continued to function in Taiwan under that name until 1963, when it was renamed the Council for International Economic Planning and Development.

In spite of the U.S. government's efforts to withdraw from direct management of scientific and technical aid to the Nationalist government, even in the 1960s and '70s, the White House continued to assist Taiwan by organizing groups of government technical experts and industrial consultants who helped the Nationalist government develop its own scientific and technical policy. For example, in 1967 President Lyndon B. Johnson sanctioned what would become known as the Hornig Mission, led by Donald F. Hornig, a chemist who was at that time serving as a presidential science adviser. The Hornig Mission consisted of a group of scientific and technical experts drawn from the public and the private sectors who traveled to Taiwan to evaluate and make recommendations for

Taiwan's scientific and technical development. The Hornig Mission ended up playing an influential role in Taiwan, where the state drew heavily on its recommendations as it developed strategies and built institutions to promote scientific and technical development in the 1970s.[38] These new institutions and the development plans they produced with input from U.S. advisers were fundamental building blocks in the broader economic development strategies that shaped the direction in the 1970s and '80s of Taiwan's developmental state. Although the people had, for the most part, changed, as had many of the institutions, some of the patterns of inter-action that had emerged during the war, particularly those involving American technical consultants, continued to function in meaningful ways for decades after the war ended.

Conclusion

Although the war caused unimaginable hardships and led to huge chal-lenges, it also created opportunities for the Nationalist state, individual Chinese technicians and scientists, and American businesses. The Nation-alist state needed to develop industry and agriculture to not only sup-port a newly enlarged population as well as a war effort, but also to sup-port postwar reconstruction of the entire nation. To accomplish these goals, it needed to acquire knowledge about the region and its resources as well as to develop human resources. As we saw in chapter 1, govern-ment technocrats, scientists, and technicians attached to state-run insti-tutions and independent scientists and technicians all worked to gather information about inland China and many used that information to in-form plans for both short- and long-term development. Chapter 2 showed us how the state also strove to take greater control over education and training with an eye to producing the human resources that would be needed to implement those plans. Chapter 4 provided insight into state efforts to gain control over industrial and infrastructural development. Taken together, these stories all show a state that took advantage of the wartime to increase its control over the hinterland through the acquisi-tion of knowledge, integration of the region into state planning for post-war development, and development of infrastructure and industry. They also show a state that was using this period to extend its control over both the educational and industrial sectors.

The Nationalist government also used the wartime period to fully commit itself to the idea of central planning. We can see this not only in the creation of a Central Planning Bureau and the development of a national plan for postwar economic development, as described in chapter 1, but also in the NRC's aggressive pursuit of contracts with U.S. firms to advance the goals articulated in that plan, described in chapter 5. This planning ethos so permeated the bureaucracy and its allies that, as we saw in chapter 3, even Chinese Americans and Chinese in America looked for ways to contribute to plans for China's postwar development. Many of these plans were for naught as the Nationalists had such limited opportunity to implement them after the end of the war. However, the case of Taiwan sugar shows that during the war, through this planning process, the NRC had positioned itself to identify important industries even in territories it did not control and lead their postwar development.

One important element of the execution of plans was acquisition of knowledge and tools. To this end, as described in chapter 3, the Nationalist government established the Universal Trading Corporation and the NRC-NY along with the Washington-based China Supply Commission, all of which worked to purchase machinery and materials and develop relationships with American businesses, universities, and government agencies. Also working with these entities were nongovernment groups such as the Committee on Wartime Planning and the Chinese Institute of Engineers, America Section, both of which maintained well-developed networks among Chinese technicians in the United States and produced publications that aimed to convey useful technical information back to technicians in China. Through these various institutions, and usually with the support of the U.S. government, the Nationalist government and its representatives promoted and facilitated a transfer of technical knowledge to China.

Another important dimension of planning was human resource development. Without adequately trained personnel, the Nationalists would not be able to carry out their plans for industrial and agricultural development. To meet this demand, the Ministry of Education worked to reorient scientific and technical education toward applied sciences and the NRC and other Nationalist agencies arranged training programs. Some of these training programs took place in China, as we saw in chapter 2, but many, as we saw in chapters 3 and 5, took place in the United States.

Trainees were typically technical staff of NRC enterprises or other government organs that supported industrial, agricultural, or military development. Most such industrial enterprises and scientific organs were newly built or rebuilt in western China in conditions that were not conducive to development. Transportation was poor, power was scarce, and raw materials were hard to come by. The industrial trainees were already well aware of the vast differences between American industrial conditions and the situation at home in China, and many of them were used to having to find workarounds to "standard" industrial procedures just to make their enterprises at all productive. Agricultural trainees were also cognizant of the very different agricultural patterns and norms in the United States as opposed to China. For as much as the training programs in the United States aimed to provide trainees with modern American industrial and agricultural know-how, therefore, most trainees, as we saw in chapters 3 and 5, returned to China with an understanding that they would have to find ways to adapt their new knowledge to a significantly different context. While in the United States, however, most trainees appear to have soaked in everything they could get from the experience and were pleased with the knowledge they were acquiring even if they knew it might not be quite suited to their situations.

American advisers to China do not all appear to have perceived these challenges and differences, however, and many of them were persistent in their expectation that China could only develop successfully if it were to follow an American path. Donald Nelson was especially outspoken on this point, demanding, as we saw in chapter 4, that China create its own War Production Board following the American model. The Nationalist government, in desperate need of American technical assistance, complied, but the CWPB never mustered the authority that the American War Production Board had, and it is quite possible that in spite of Weng Wenhao's protestations to the contrary, no one in the Chinese government ever really expected that it would. Although some American advisers attached to the AWPM were cut of the same cloth as Nelson, others came to a much better understanding of the challenges that Chinese industry faced and expressed their admiration for the resilience and creativity of the Chinese technicians they encountered in the industries that they advised.

In the waning months of the war and after its end, the NRC increasingly looked to hire its own technical consultants and, in doing so, to take greater control of the implementation of its own development plans. American technical consultants often had their own agendas, however, and were eager to develop relationships with the Chinese government that could benefit the bottom lines of their companies. As we saw in chapter 4, some of the AWPM consultants were motivated by these interests, but chapter 5 demonstrated that such motives were even more apparent among the independent industrial consultants hired by the NRC, even those hired while the war was still ongoing. The words and behavior of these consultants make it glaringly apparent that postwar economic opportunity was just as important to Americans working with China as it was to the Nationalist government.

Following the war, the U.S. government extracted itself from its wartime pattern of providing technical assistance to China and embraced the idea that such efforts should take place as business transactions in the private sector. For a time, this appeared to work for the Nationalists, who, as the LSU sugar training program demonstrates, appear sometimes to have manipulated that system to their own advantage. However, the Chinese could only sometimes afford to pay for these contractual services themselves and were dependent on U.S. loans to support other contracts, loans that were not always forthcoming. Under these circumstances, even though the Nationalist government may have had somewhat more agency in determining which kinds of projects to develop than they had had under the CWPB-AWPM model, they were only able to realize their own plans in cases where they could afford to pay for the consultants and the plans the consultants came up with. The refusal of U.S. lending agencies to support other plans meant that those plans could not be realized.

Although many of these wartime development efforts, particularly the technical training efforts, laid an infrastructural and human resources foundation for continued industrial and agricultural development in China, their legacy can be seen more clearly in Taiwan. Many of the bureaucrats and technicians who had focused their attention on the development of one hinterland region of China during the war ended up turning their sights on a new peripheral region as reconstruction of Japanese-occupied areas of the mainland became increasingly impossible. In Taiwan, both before and after 1949, the NRC and its successor institutions continued to utilize the technical and organizational knowledge along with the connections and international advising infrastructure they had developed during the war. In China, although the knowledge continued to be used, most of those connections were suspended until the late 1970s, when former students and trainees who had spent time in the United States during the war began to revive them. Some of the former trainees did not survive the tumultuous years of the 1960s, but others continued to contribute to China's industrial, military, and agricultural development for decades after the establishment of the PRC.

Notes

Introduction

1. Buck, *American Science in Modern China*, 224.

2. Agricultural scientists appear to have been the main exception to this rule. Xuan Geng has shown in her dissertation, "Serving China through Agricultural Science," that quite a number of elite Chinese agricultural scientists were highly motivated by a nationalistic desire to improve China.

3. Lockhart, "Memorandum on Points of Possible Discussion with Mr. T. V. Soong." See also "Notes from Mr. Pierce," 1.

4. Lean, *Vernacular Industrialism in China*, 14.

5. Tang, *Science and Technology in China*, 239.

6. MacNair, *Voices from Unoccupied China*, xliv–xlix. For a fuller discussion of divergent opinions on the merits of pure versus applied sciences in China during this period, see Fu, "Science, Society and Planning," 112–113 and Fu, *Plan of Science and Planning of Science in Republican China*.

7. Reardon-Anderson, *Study of Change*, 314–318.

8. Tan, *Recharging China in War and Revolution*, ch. 3.

9. Lepore, "Our Own Devices," 90.

10. In addition to Buck and Reardon-Anderson cited above, see, for example, Bullock, *Oil Prince's Legacy*; Lei, *Neither Donkey nor Horse*; Schneider, *Biology and Revolution*, 2003; Hu, "American Influence on Chinese Physics Study"; Hu, "Reception of Relativity in China."

11. Shen, "Murky Waters," 587.

12. See, for example, Lowdermilk, "Soil, Forest, and Water Conservation," 408–409.

13. Fairbank, *America's Cultural Experiment*, 26–27, 59–60; "The Ambassador in China (Gauss) to the Secretary of State," December 9, 1943. Burt also discusses this issue in chapter 5 of *At the President's Pleasure*.

14. See, for example, "Summary of Conversation Held in Chungking on September 12, 1944, between Dr. Wong Wen-Hao Minister of Economic Affairs, and Mr. Nelson";

"Glen M. Ruby to C. D. Shiah"; "Notes from Mr. Pierce"; Locke, "Report to the President," 304–316; Hanson, "Blue Print for Exporting American Know-How."

15. Marx, "Technology," 575.

16. Edgerton, *Shock of the Old*.

17. Downey and Lucena, in "Knowledge and Professional Identity in Engineering," discuss how engineers in France, Germany, the United Kingdom, and the United States redefined the contents of engineering education and knowledge in different contexts so that engineering had different meanings in different times and places. They also argue that engineers code switch between different "codes of meaning" across national boundaries. These ideas, in combination with Vanessa Ogle's observations about the adoption of international standard time in Bombay and Beirut, are helpful for understanding the interplay between the global and the local and how that interplay can be shaped by and shape national identity. Ogle, "Whose Time Is It?"

18. Surman, "Circulation, Globalization, Go-Betweens," 15–18.

19. Surman, "Circulation, Globalization, Go-Betweens," 15; Raj, "Beyond Postcolonialism," 344.

20. Shen, *Unearthing the Nation*, 16.

21. Xue, *Thirty-One Scholars of the War of Resistance*, 87–107.

22. Bullock, *Oil Prince's Legacy*; Lei, *Neither Donkey nor Horse*; Soon, *Global Medicine in China*.

23. Paul B. Trescott has done meticulous research on educational background, job placement, and research output of Chinese economists in the first fifty years of the twentieth century and has found that not only were the vast majority of Chinese economists of this period trained in the United States, but U.S.-trained economists also accounted for the greatest research output and dominated China's major centers for economic research. See Trescott, *Jingji Xue*.

24. See Bieler, *"Patriots" or "Traitors"?* For more on Chinese students in the United States in the early to mid-twentieth century, see Ni, *Cultural Experiences of Chinese Students*.

25. For studies of academic networks, see Trescott, Buck, and Chiang, *Social Engineering and the Social Sciences in China*. For material on friends of China, see Bullock, *Oil Prince's Legacy*, and several of the chapters in Ryan, Chen, and Saich, *Philanthropy for Health in China*. A number of activist friends of China have written their own accounts of their activities in China during the war. These include works describing various dimensions of the Gung Ho movement, such as Alley, *Sandan*. Materials on the U.S.-China relationship during the war have generally focused on the military and political relationship, such as Tuchman, *Stilwell and the American Experience in China*, Tsou, *America's Failure in China*, and May, *Truman Administration and China*.

26. Wong, *Americans First*.

27. Hsu, *Good Immigrants*.

28. Fairbank, *America's Cultural Experiment*; Gragg, "History of the American War Production Mission in China."

29. Kirby, "Continuity and Change in Modern China," 123; Esherick, "Ten Theses on the Chinese Revolution," 48.

30. Bian, "Building State Structure."

31. Bian, "Redefining the Chinese Revolution," 316.

32. Kinzley, "Crisis and the Development of China's Southwestern Periphery."
33. Tan, *Recharging China in War and Revolution.*
34. Lecuyer, "Making of a Science Based Technological University."
35. Lindee, *Rational Fog,* 19.

Chapter 1

1. Bian, "Redefining the Chinese Revolution"; Kirby, "Continuity and Change"; Zanasi, *Saving the Nation.*
2. "Mr. K. N. Chang's Speech."
3. "C.I.E. Forum on Post-war Industrialization of China," 3. In 1943, most of the members of the CIE were Chinese who were either studying abroad or working in the United States and who could not easily return to China.
4. See Shen, *Unearthing the Nation,* 125–127; Eastman, *Abortive Revolution*; Kirby, "Engineering China."
5. Shen, *Autobiography of a Chinese Farmer's Servant,* 151.
6. National Bureau of Industrial Research, "NBIR Work Plan."
7. Hostetler, *Qing Colonial Enterprise*; Fan, *British Naturalists in Qing China*; Stein, *Memoir on Maps of Chinese Turkistan and Kansu*; Glover et al., *Explorers & Scientists in China's Borderlands, 1880–1950*; Lam, *Passion for Facts*; Mueggler, *Paper Road.*
8. See Weller, *Caravan across China.*
9. Various articles in *Dili xuebao,* vols. 1–4.
10. Swen and Chu, *Production and Marketing of Wood Oil.* Although the survey was not published until 1942, it is clear from the text that the study was written in 1939 and not updated to reflect the changing realities of the early 1940s.
11. Shen, *Agricultural Resources of China,* 252.
12. Zhou, Hou, and Chen, *Economical Atlas of Sichuan.*
13. Bao and Chang, "The Land Utilization of the Lo-Lung River Basin," English abstract, 8.
14. Jen et al., "Land Utilization in Tsunyi District, Kweichow," 1.
15. Jen et al., "Land Utilization in Tsunyi District, Kweichow," 13.
16. Jen et al., "Land Utilization in Tsunyi District, Kweichow," 13–14.
17. Shen, *Unearthing the Nation,* 127.
18. Shen, *Unearthing the Nation,* 148.
19. Tseng, "Open Letter to British Scientists," 5.
20. Tseng, "Open Letter to British Scientists," 2, 5.
21. Lee, "Report of the Northwestern China Scientific Expedition," 1–30.
22. Hao, "Southwest Gansu's Forest," 64.
23. Zhan, Wang, and Deng, "The General View about the China Institute of Geography." The China Institute of Geography later morphed into the Chinese Academy of Science Preparatory Office.
24. Shen, *Unearthing the Nation,* 158–159; Northwest Branch of the National Geological Survey, *Overview of the Northwest Branch,* 2. As Grace Shen has argued, the China Geological Survey's Northwest Branch developed a model for regional fieldwork that was later employed in both Taiwan and Manchuria.

25. Tseng, "Open Letter to British Scientists," 5.

26. "Expeditions of Sinkiang," 6; "Academia Sinica's Northwest Scientific Investigation Team Cost, Budget, and Unit Work Plans, Etc." The Academia Sinica files indicate that a great deal of preparatory work and financial outlay had to go into these expeditions. The expedition projected costs ahead of their trip, but documents show that the total cost ended up being more than anticipated and members of the expedition had to keep requesting additional funding to support the trip. Moreover, local governments also requested support from the central government to cover the costs of hosting the expedition. For example, one county government in Xinjiang sought reimbursement for the costs of repairing the expedition's car. The researchers were aided in their work by a cook, two drivers, two bodyguards, and an interpreter, although the documents do not clarify which regional language(s) the interpreter spoke. They took a car plus supplies for the car and for themselves, but they also rode horses when accessing areas that cars could not reach. They also took an extensive supply of medicines, including quinine and aspirin. The file shows that the Executive Yuan, the Ministry of Education, and the Supreme Military Council were all interested in their work. In fact, they had enough clout to get an order permitting them to be flown to Xinjiang to investigate the northern border areas.

27. Ho, "Reminiscences," 337.

28. Ho, "Reminiscences," 338.

29. See, for example, Weller, *Caravan across China*, 234–236; Dykstra to wife, October 15, 1943; Needham, Friday 20th August, "Journal of North-West Tour."

30. Ho, "Reminiscences," 351.

31. Ho, "Reminiscences," 350.

32. "Linking the Northwest and the Southwest," 37. Chen was reputedly, along with Tao Xisheng, a coauthor of Chiang Kaishek's 1943 volume *Chinese Economic Theory*.

33. Ho, "Reminiscences," 338.

34. Ho, "Reminiscences," 352–353.

35. Luo, *Report on the Northwest Expedition*, 1–7.

36. Luo, *Report on the Northwest Expedition*, 34.

37. Luo, *Report on the Northwest Expedition*, 42–48.

38. Luo, *Report on the Northwest Expedition*, 57–58.

39. Zhou et al., *Economical Atlas of Sichuan*, 2.

40. Graham, *Toward a Planned Society*, 9.

41. Fu, "Science, Society and Planning," 102.

42. Chiang, as quoted in Fu, *Plan of Science and Planning of Science*, 90.

43. Chinese Ministry of Information, *China Handbook, 1937–1945*, 96.

44. Wu, "Economic Reconstruction and Planning: Wartime and Post-war," in MacNair, ed., *Voices from Unoccupied China*, 72–75.

45. Chinese Ministry of Information, *China Handbook, 1937–1945*, 97.

46. Bian, *Making of the State Enterprise System in Modern China*, 162.

47. Ho, "Reminiscences," 355.

48. Ho, "Reminiscences," 357–362.

49. Perhaps not coincidentally, this topic was also hotly debated outside of government circles. See, for example, Chen, "Some Outstanding Problems."

50. Central Planning Bureau, *Five Year Plan for China's Economic Development*, 1–6; Ho, "Reminiscences," 363–374. The KMT did, however, use the plan to guide at least some of its postwar development efforts and provided it, or parts of it, to contractors it hired to assist with these efforts in 1945 and 1946 (see chapter 5).

51. Sumner, "Chinese Attitudes toward Postwar Economic Policy," 5.

52. Peng, "A Program for Sheep Improvement."

53. Peng, "A Program for Sheep Improvement."

54. Chinese Ministry of Information, *China Handbook, 1937–1945*, 418.

55. See, for example, Zhang, "Livestock of Southwest Gansu," 87–89; Phillips, Johnson, and Moyer, *Livestock of China*.

56. Phillips et al., *Livestock of China*, 2, 91.

57. See, for example, Jiang, "Final Report," or Liang, "Chinese Training Program," both discussed in chapter 5.

58. Tsou and Chang, "Program for Postwar Agricultural Reconstruction in China," 1–2.

59. Tsou and Chang, "Program for Postwar Agricultural Reconstruction in China," 6–8.

60. Tsou and Chang, "Program for Postwar Agricultural Reconstruction in China," 9–21.

61. Tsou and Chang, "Program for Postwar Agricultural Reconstruction in China," 22.

62. Pan, "Plan Proposed for Improving Alcohol Industry in China," 2.

63. Pan, "Plan Proposed for Improving Alcohol Industry in China," 16–21.

64. Pan, "Plan Proposed for Improving Alcohol Industry in China," 35–36.

65. Meng, *Chinese American Understanding*, 188–189.

66. "Report of the Midwestern Conference," 159. *National Reconstruction Journal* was the journal of the Committee on Wartime Planning.

67. Hou, "Nationalization of Chinese Heavy-Chemical Industries," 19; Li, "Post-war Industrialization Problems" 31; Chen, "Some Outstanding Problems," 38.

68. "Summary of Resolution on Industrial Reconstruction Principles," 2. J. B. Condliffe, a Berkeley economist and member of the Institute of Pacific Relations, fearing for China's ability to preserve its independence from colonial masters, argued strenuously that China should not rely on foreign capital to develop after the war, whereas Li Ming, as noted earlier, felt that foreign investment was an opportunity that both the government and entrepreneurs should take advantage of. See Condliffe, "The Industrial Development of China," 73.

69. "C.I.E. Forum on Post-war Industrialization of China," 3.

70. Chen, "Some Outstanding Problems," 38–39.

71. The Chinese Economic Reconstruction Society was organized in 1938 and included "bankers, merchants, manufacturers, engineers, government officials, university professors, research workers and other professional people in a wide variety of fields." It produced a "Draft Outline of the Principles for China's Post-war Economic Reconstruction" in October 1940. Fong, Lin, and Koh, "Problems of Economic Reconstruction in China," 1.

72. See, for example, Pu, "A Brief Report on the TVA," 3–6. Pu was a Chinese engineer who had been sent by the NRC to train at and study the Tennessee Valley Authority.

73. Tsou and Chang, "A Program for Postwar Agricultural Reconstruction in China," i (prologue).

Chapter 2

1. Li, "Reminiscences of Li Hanhun," 129.

2. Hu, as quoted in Fairbank, *America's Cultural Experiment*, 11.

3. Chin, as quoted in MacNair, *Voices from Unoccupied China*, l–li.

4. Tsai, as quoted in MacNair, *Voices from Unoccupied China*, li.

5. A search of *rencai* (人才) in the Academia Historica online catalog reveals about 650 results from the period between 1927 and 1950. Of these, many are related to the topic of training or developing various sorts of *rencai*, which can be translated as "talent" but which is also used to refer to "personnel." The term was not used exclusively for scientific and technical personnel. Therefore, some of these documents are addressing the need for or plans to develop other kinds of talent as well.

6. See Luo, *Report on the Northwest Reconstruction Expedition*, 57–58.

7. Central Planning Bureau, "First Five-Year Program for China's Postwar Economic Development," 26–27 or "Draft Five-Year Program for Economic Development," Overview, Table 5.

8. Luo, *Report on the Northwest Expedition*, 409.

9. Luo, *Report on the Northwest Expedition*, 394–395.

10. Luo, *Report on the Northwest Expedition*, 177–179.

11. Chen, as quoted in Freyn, *Chinese Education in the War*, 93. The Chen Lifu article appeared in the *Shao Tang Pao*, Hankow, July 7, 1938.

12. Chen, *Chinese Education in the War*, 4. Parts of this chapter draw on work I previously published in "Looking Toward the Future."

13. Chen, *Chinese Education in the War*, 4.

14. Yang, "Chinese Education, Past and Present," 114.

15. Ministry of Education, *University Course Goals*, 4–7.

16. "A Working Program for the Improvement of Agricultural Education in China," 1–2.

17. Yang, "Chinese Education, Past and Present," 112.

18. Fairbank, *America's Cultural Experiment*, 29.

19. Fairbank, *Chinese Educational Needs and Programs*, 17.

20. Freyn, *Chinese Education in the War*, 124–125.

21. Freyn, *Chinese Education in the War*, 126.

22. Chen, *Storm Clouds Clear*, 169–170.

23. Chen, *Storm Clouds Clear*, 169–170.

24. China Information Committee, *China after Four Years of War*, 126.

25. "Fazhan guofang kexue."

26. See Schmalzer, *The People's Peking Man*, 29–31, 47–49; Lean, *Vernacular Industrialism*, 56, 77–119.

27. "Rules for County and City Science Organs," 7–21.

28. Nanjing Second Historical Archive, Ministry of Education Archives, 5-12244, 5-12249, 5-12265.

29. Joseph Needham and Dorothy Needham, C. 56, C. 60, C. 63, Needham Archives, University Library, Cambridge.

30. Needham, "Science and Technology in the North-West of China," 240.

31. "A Newspaper on Popular Science Displayed outside the Kansu Science Education Institute," NW1–22. Needham Research Institute collection of wartime photos, http://www.nri.org.uk/JN_wartime_photos/nw.htm.

32. Nanjing Second Historical Archive, Ministry of Education Archives, 5-12265, 5-12280, 5-12282.

33. Needham, "Chungking Industrial and Mining Exhibition," 672. See also National Resources Commission, *Guide to Industrial & Mining Exhibition*; Taylor, "I Attend a Chungking Exhibition," 64–68. Taylor had been sent to China to do journalism training as part of the U.S. State Department's cultural relations program (see chapter 3).

34. British Council, "Industrial and Mining Exhibition," 7.

35. Needham, "Chungking Industrial and Mining Exhibition," 675.

36. Nanjing Second Historical Archive, Ministry of Education Archives, 5-12280, 5-12282.

37. Hu, *"Childhood."*

38. Joseph Needham and Dorothy Needham, C. 66, Needham Archives, University Library, Cambridge.

39. "Fazhan guofang kexue."

40. "Telegram from Chiang Kaishek to the Ministry of Education," 617–618. In 1942, the Science and Technology Planning Committee was reorganized into the National Defense Science and Technology Planning and Improvement Committee. This is just one of a long series of telegrams about the creation of this committee.

41. Yeh, *Wartime Shanghai*, 36–40.

42. In fact, the MOAF was reorganized several times over the course of the war so as to better structure/streamline the relationship between these diverse organizations.

43. "Ma Baozhi." Ma, who also went by the English name of Paul C. Ma, went on to become the dean of National Taiwan University's School of Agriculture from 1954 to 1961 and also later ran programs to train agricultural technicians at the University of Liberia. According to Hanson et al., "Introducing Agricultural Engineering in China," vii, after the war, Ma became director of the agricultural division of the MOAF and director of China's Agricultural Machinery Operation and Management Office.

44. Needham, "Chinese Papers, 1942–1946," Vol. 1, "Report of a Journey in the South East of China, Occupying April, May and June, 1944," 52–53; Ma Baozhi and the Liuzhou Animal Husbandry School.

45. "A New System of Schools," 2–3.

46. Ministry of Economic Affairs, "Report of the Works of the Ministry of Economic Affairs, August 1944," 8.

47. Ho, "Reminiscences," 194, 221–222.

48. Ho, "Reminiscences," 235–236.

49. Ho, "Reminiscences," 235.

50. Ho, "Reminiscences," 226–227.

51. Ho, "Reminiscences," 236–237.

52. See documents in Nanking Academic, Related to Faculty and Staff, RG011-198-3399.

53. Hogg, "Bailie Schools," 2–5.

54. International Committee for Chinese Industrial Cooperatives, "Shantan Bailie School Training School for the Chinese Industrial Co-operatives," 504.

55. "George Hogg to Ida Pruitt, June 21, 1942."

56. Hogg, "Bailie Schools," 1.

57. Hogg, "Plan for Lanchow Bailie School."

58. Hogg, "Report on the Shuangshihpu Bailie School," 2, 11.

59. International Committee for Chinese Industrial Cooperatives, "Shantan Bailie School," 504; Alley, *Sandan*, 17–18.

60. Hogg, "Report on the Shuangshihpu Bailie School," 6.

61. Hogg, "Report on the Shuangshihpu Bailie School," 7.

62. Hogg, "Letter to Ida Pruitt," July 23, 1944, 6.

63. Hogg, "Bailie Schools," 3.

64. Hogg, "Report on the Shuangshihpu Bailie School," 14–17.

65. Alley, "Shantan Bailie School," 8.

66. Alley, "Shantan Bailie School," 10.

67. Alley, "CIC in the Northwest," 2.

68. Alley, "CIC in the Northwest," 2.

69. Needham, "Report to His Excellency President and Generalissimo Chiang Kai-Shek on the Position and Prospects of Science and Technology in China," in "Chinese Papers, 1942–1946," 2:55. Needham is referring to the Liuchow Agricultural Vocational Middle School.

70. "Needham Complete Diaries," NRI, entries for Wednesday, August 18, and Monday, November 8, 1943.

71. Needham "Letter to Sir Ronald," October 23, 1947.

72. Pruitt herself was born and raised in rural Shandong Province, the daughter of American missionaries. She traveled to the United States for education and returned to China first to teach in a girls' school and later, following graduate work, as head of the Department of Social Services of the Rockefeller Foundation–run Peking Union Medical College.

73. Barnett, "China's Industrial Cooperatives," 51–56.

74. Fairbank, *America's Cultural Experiment*, 5.

75. Stuart E. Grummon, head of the Division of Cultural Relations, as quoted in Fairbank, *America's Cultural Experiment*, 15.

76. Fairbank, *America's Cultural Experiment*, 18; "The Ambassador in China (Gauss) to the Secretary of State," March 27, 1942, 706–707. The Chinese government also requested material (microscopes, lab equipment, and books) and funding to augment higher education in China.

77. "Letter from Chen Lifu to Jiang Tingfu on the List of American Experts Needed by the Ministry of Education," 3, in He, *Documentary Collection*; "The President of the Chinese Executive Yuan and Acting Minister for Foreign Affairs (Chiang Kai-shek) to the American Ambassador (Gauss)," June 20, 1942, 713.

78. Nanjing Second Historical Archive, Academia Sinica Archives, 393–365.

79. "Letter from Y. T. Ku to William H. Dennis," Nanjing Second Historical Archive, 448–543.

80. "Memorandum Prepared in the Department of State," 715.

81. Fairbank, *America's Cultural Experiment*, 57.

82. "Memorandum by Mr. Haldore Hanson of the Division of Cultural Relations," July 15, 1942, FRUS, China, 1942, p. 718.

83. "The Acting Secretary of State to the Ambassador in China (Gauss)," October 3, 1942, 723.

84. Needham, "Memorandum on Western Technical Experts in China." The same file also contains comments on Needham's memo by various Foreign Office and British Council bureaucrats. As chapters 4 and 5 show, Needham's Foreign Office critics were not entirely wrong in their assessment of American motives.

85. British Council, "Dr. Needham's Memorandum."

86. "Mr. Eden to Sir H. Seymour," July 27, 1943, in BDFA, Vol. 7, Part 23, 37; "Sir H. Seymour to Mr. Eden," August 14, 1943, BDFA, Vol. 7, Part 23, 47; "Sir H. Seymour to Mr. Eden," September 15, 1943, BDFA, Vol. 7, Part 23, 77.

87. K. T. Gurney comments on "Technical Experts for China"; A. Scott, comments on "Technical Experts for China."

88. A. Scott comments on "Technical Experts for China."

89. "Sir H. Seymour to Mr. Eden," October 4, 1944, BDFA, Vol. 7, Part 26, 123.

90. Fairbank, *America's Cultural Experiment*, 3.

91. Dykstra, Letter to His Wife, October 15, 1943, Dykstra Papers.

92. Dykstra Diary, August 12, 1943, April 13, 1944, Dykstra Papers.

93. Dykstra, Letter to His Wife, December 28, 1943 and January 3, 1944, Dykstra Papers.

94. Dykstra Diary, January 4, 1944, March 7, 1944 and Dykstra, Letter to His Wife, March 11, 1943, Dykstra Papers.

95. Needham, "Memorandum on Western Technical Experts," 4.

96. Needham, "Memorandum on Western Technical Experts," 6. The three oil drillers—Bush, Field, and Reiner—to whom Needham was referring, were not, in fact, part of the State Department program. One petroleum geologist, Ed Beltz of Standard Oil, did go to Yumen as a State Department expert, but he did not remain there as long as the other three.

97. Dykstra, Letter to His Wife, February 16, 1943, Dykstra Papers.

98. Dykstra, Letter to His Wife, July 24, 1943 and Dykstra, Letter to His Wife, November 26, 1943, Dykstra Papers.

99. Dykstra Diary, September 18, 1943, Dykstra Papers.

100. Dykstra Diary, January 27, 1944, Dykstra Papers.

101. Ministry of Agriculture and Forestry, "Report of the National Agricultural Research Bureau," 27–28.

102. Dykstra tells a detailed story of one such interaction with a farmer in which it is clear that Chang and Yang had to do a great deal of negotiating with the farmer to get him on board. Dykstra, Letter to His Wife, May 4, 1944, Dykstra Papers.

103. Dykstra, Letter to His Wife, October 15, 1943, Dykstra Papers.

104. Lowdermilk, "Preliminary Report to the Executive Yuan," 5.

105. Lowdermilk, "Preliminary Report to the Executive Yuan," 4.

106. Lowdermilk, "Soil, Forest, and Water," 392.

107. Lowdermilk, "Soil, Forest, and Water," 409, 393.

108. Lowdermilk, "Soil, Forest, and Water," 393; "Economic Milestones."

109. Lowdermilk, "Soil, Forest, and Water," 408–409.

110. Fan, *British Naturalists*, 147.

111. Lowdermilk, "Soil, Forest, and Water," 388, 392, 395; Lowdermilk, "Preliminary Report to the Executive Yuan"; Lowdermilk, "Statement for News Notes for Chinese Students," 2–4.

112. Lowdermilk, "Soil, Forest, and Water," 392.

113. "W. H. Wu to L. F. Chen," September 16, 1943.

114. Eaton, "China's Industrial Training Program," 206.

115. Eaton, "China's Industrial Training Program," 207.

116. Eaton, "China's Industrial Training Program," 206–207.

117. "L. F. Chen to Paul Eaton," May 11, 1945.

118. *CIE News Bulletin* 5, no. 1 (February 10, 1946): 3. See also earlier references to Eaton in other issues of the journal; "L. F. Chen to Eaton," May 10, 1945.

119. Lowdermilk, "Soil, Forest, and Water," 393.

Chapter 3

1. Although early twentieth-century Chinese students going abroad to study on Boxer fellowships were mandated to focus on applied learning, other Chinese students were not subject to such restrictions. The Nationalists' interventions thus represented a shift from past practice.

2. "The Ambassador in China (Gauss) to the Secretary of State," June 3, 1944, 1144.

3. "Memorandum of Conversation, by the Ambassador in China (Gauss)," April 12, 1944, 1135–1136; see also "Memorandum of Conversation, by the Special Assistant in the Division of Science, Education and Art (Peck)," 1140–1142 for a discussion of whether the "superintendent of students" position was intended to implement thought control or whether it was simply meant to be a supervisor with whom institutions of higher education in the United States could communicate about student behavior and performance.

4. "The Ambassador in China (Gauss) to the Secretary of State," May 4, 1944, 1139–1140.

5. "The Ambassador in China (Gauss) to the Secretary of State," September 24, 1944, 1152; "The Ambassador in China (Gauss) to the Secretary of State," May 30, 1944, 1143.

6. Meng, "Training of Chinese Technical Personnel," 33.

7. Kwoh, "Chinese Students in American Universities," 9.

8. Kwoh, "Chinese Students in American Universities," 12.

9. Li, *Freedom & Enlightenment*, 55.

10. Nanjing Second Historical Archive, National Industrial Research Bureau Archives, 448–543.

11. Meng, "Training of Chinese Technical Personnel," 32–33.

12. Meng, "Training of Chinese Technical Personnel," 34.

13. Meng, *Chinese American Understanding*, 186–188.

14. Meng, *Chinese American Understanding*, 186.

15. "Announcing Fellowships for Graduate Chinese Students."

16. "China Institute in America, 60th Anniversary," 37; Meng, *Chinese American Understanding*, 183.

17. "Chinese Studying in American Colleges," 1.

18. Meng, *Chinese American Understanding*, 186. For more on the Committee on Wartime Planning, see chapter 1.

19. "Organizational Report," 107.

20. "Organizational Report," 105.

21. "Organizational Report," 107.

22. "Report of the Midwestern Conference," 157.

23. Meng, *Chinese American Understanding*, 215.

24. Ni, *Cultural Experiences of Chinese Students*, 145.

25. Meng, "Training of Chinese Technical Personnel," 34–35.

26. Meng, "National Reconstruction Forums," 1–2.

27. Meng, "National Reconstruction Forums," 2.

28. Meng, "Training of Chinese Technical Personnel," 34.

29. Meng, "Foreword," *National Reconstruction* 5 (July 1944): iv.

30. Meng, "Foreword," *National Reconstruction* 4 (February 1944): 1.

31. For example, in *Recharging China in War and Revolution*, Ying Jia Tan mentions several individual technicians working in electrical fields whom the NRC sent to the United States, Britain, and elsewhere prior to the war.

32. Meng, "National Reconstruction Forums," 2.

33. Meng, "Committee on Wartime Planning," 5.

34. Meng, "Chinese-American Educational," 2.

35. "Chia Wei Chang to Lowdermilk"; "Chia Wei Chang to Chih Meng."

36. "Notes on March 13, 1942 Meeting"; "Lowdermilk to H. G. Calkins."

37. "Lowdermilk to Chia Wei Chang"; "Lowdermilk to D. Y. Lin."

38. "Lowdermilk to Chih Meng."

39. Meng, "China Institute," 36–37.

40. Meng, "China Institute," 37.

41. For information on one factory program, see Shih, *China Enters the Machine Age*, 148–149.

42. *CIE News Bulletin* 1, no. 1 (October 10, 1942).

43. *CIE News Bulletin* 1, no. 2 (December 1942), 2.

44. "Foreword," *CIE Journal* 1, no. 1 (April 1943): 2.

45. "Heard at the Inauguration Meeting," 5.

46. "Announcement of Opportunities" and "Chinese on the Job"; "Positions Open." As noted in a listing for Portland, Oregon's Electric Steel Foundry Company in "Employment Opportunities," "The company expects to do business in China after the war and they would like those engineers trained with them to work for them later, but they do not make this a contractual obligation."

47. "Recruiting Technical Manpower."

48. "The Department of State to the British Embassy."

49. The loan was negotiated prior to the U.S. entry into the war, and the United States was particularly concerned that it not send the wrong signal to Japan. Haw, "Tung Oil and Tong 桐Trees," 218; Yu, "Tung Oil Production and Its World Market," 36; "Universal Trading Corporation," 105–108, 134, 137; Zheng, *Research on the Nationalist Government's Wartime Control of Economy and Trade*, 232.

50. "Foreword," *Universal Engineering Digest* 1, no. 1 (April 1943): 1.

51. See Tan, *Recharging China in War and Revolution*, chapter 3 for a discussion of negotiations on a technology transfer agreement between the NRC and Westinghouse that ran through the UTC.

52. "Foreword," *Universal Engineering Digest* 1, no. 1 (April 1943): 1.

53. See Kwan, *Patriot's Game*. Yongli still operates as a state-owned enterprise in Tianjin.

54. "Hou Te-Bang." For more information on the Yongli Chemical Industries, see Kwan's detailed study, *Patriot's Game*.

55. Hou, "Nationalization of Chinese Heavy-Chemical Industries," 13–19.

56. Meng, *Chinese American Understanding*, 189; *CIE News Bulletin* 3, no. 6 (December 10, 1944): 3.

57. *CIE News Bulletin* 3, no. 1 (February 10, 1944): 4. As seen in chapter 1, Tsou had also been a member of the CPB's Northwest Reconstruction Investigation Team.

58. "Regulations Governing the NRC Technical Group in USA."

59. "Letter to Dr. Ku and Co-workers from Wm. Han-chu Lee et al."

60. *CIE News Bulletin* 3, no. 1 (February 10, 1944): 3; *CIE News Bulletin* 2, no. 3 (June 10, 1943).

61. "Foreword," *Universal Engineering Digest* 1, no. 1 (April 1943): 1.

62. *Universal Engineering Digest* 1, no. 3 (August 1943).

63. Cheng and Cheng, *Archives on the National Resources Commission Technicians' Training in the United States*, 1:9.

64. Cheng and Cheng, *Archives on the National Resources Commission Technicians' Training in the United States*, 1:11.

65. Cheng and Cheng, *Archives on the National Resources Commission Technicians' Training in the United States* 1:213–215.

66. Cheng and Cheng, *Archives on the National Resources Commission Technicians' Training in the United States* 1:19. Information about training destinations for all of the trainees is on pp. 12–29.

67. Cheng and Cheng, *Archives on the National Resources Commission Technicians' Training in the United States* 1:204–205.

68. "Letter from L. F. Chen to Y. S. Sun," June 22, 1944, in Cheng and Cheng, *Archives on the National Resources Commission Technicians' Training in the United States* 1:242–243; see pp. 226–259 for the full correspondence between Sun, Chen, and others related to Sun's training. Sun went on to serve as minister of economic affairs in Taiwan in the 1970s and served as premier of the ROC from 1978–1984.

69. "L. F. Chen to C. C. Wei," April 16, 1943, in Cheng and Cheng, *Archives on the National Resources Commission Technicians' Training in the United States* 1:321.

70. "Sun Jianchu to Chen Liangfu," December 7, 1943, in Cheng and Cheng, *Archives on the National Resources Commission Technicians' Training in the United States* 1:312.

71. K. Y. Yin was himself an engineer who as an employee of the Ministry of Transport had developed a close relationship with T. V. Soong. In the 1950s, he became minister of economic affairs and in that role and others, became one of the main architects of the KMT's economic development policies in Taiwan.

72. For more on the Yangzi Valley Administration project, see Yin, "The Long Quest for Greatness."

73. Cheng and Cheng, *Archives on the National Resources Commission Technicians' Training in the United States* 1:388–437.

74. Cheng and Cheng, *Archives on the National Resources Commission Technicians' Training in the United States* 1:413–414.

75. Cheng and Cheng, *Archives on the National Resources Commission Technicians' Training in the United States* 1:442–446.

76. "L. F. Chen to Y. S. Sun," March 30, 1944, in Cheng and Cheng, *Archives on the National Resources Commission Technicians' Training in the United States* 1:451.

77. Cheng and Cheng, *Archives on the National Resources Commission Technicians' Training in the United States* 2 and 3.

78. *CIE News Bulletin* 3, no. 3 (June 10, 1944): 5.

79. Chen, *Report to Annual Convention of the Technical Committee of the NRC*.

80. "Atcheson (Chargé in China) to Sec State," 744–745.

81. "Interview with Dr. Wong Wen Hao," 2.

82. See, for example, "The Ambassador to China (Gauss) to the Secretary of State," October 7, 1943, 750–751.

83. "Letter from Patrick J. Hurley to T. V. Soong"; "Notes on Conference Held with Dr. T. V. Soong," 3–4.

84. *CIE News Bulletin* 4, no. 1 (February 10, 1945): 2.

85. "Statement Concerning Administrative Costs," 2. See also "May 22, 1945 Agreement."

86. The Chinese military also sent other groups of trainees to the United States. For example, in 1944, a group of Chinese naval officers were stationed first at the Philadelphia Naval Yard and later at MIT to take a course on naval architecture and engineering. See *CIE News Bulletin* 3, no. 4 (August 10, 1944): 10.

87. Li, *Freedom & Enlightenment*, 52. Jianqiao was the first Chinese-U.S. joint venture in aviation.

88. Li, *Freedom & Enlightenment*, 64.

89. Li, *Freedom & Enlightenment*, 74.

90. Li, *Freedom & Enlightenment*, 78–82.

91. Li, *Freedom & Enlightenment*, 90.

92. Liu, "Exploring the Secrets of Aircraft Engine Factory," 109–115.

93. Wu, *Wu Daguan's Oral Autobiography*, 18–28.

94. Tsou, "A Fundamental Problem," 5–6.

95. Tsou, "A Fundamental Problem," 9.

96. Tsou, "A Fundamental Problem," 10.

97. Yang, "Fundamental Problems," 34.

98. Yui, "China's Tung Oil Export," 120.

99. "More Chinese Students Arrive"; "Chinese Scholarship Winners Will Study at Iowa State."

100. "Graff, E. F. Letter to W. H. Stacy."

101. Stacy, "Developing Voluntary Leadership," 1–2.

102. "Check List for Use."

103. "Graff, E. F. Letter to W. H. Stacy et al.," May 7, 1947.

104. Shih, "Term Paper," 1–2.

105. Shih, "Term Paper," 9.

106. Shih, "Term Paper," 7.

107. Gross, "Aid to China's Agriculture," 18–21; International Harvester, "For Release Immediately."

108. International Harvester, "For Release Immediately," 2.

109. International Harvester, "For Release Immediately," 2.

110. Hansen et al., "Introducing Agricultural Engineering in China," 143.

111. International Harvester News Releases: 1947, 1–2.

Chapter 4

1. See Tan, *Recharging China in War and Revolution.*

2. See Kwan, *Patriot's Game.*

3. "Donald M. Nelson Report to the President, December 20, 1944," in Mabel Gragg, "History of the American War Production Mission in China," 218.

4. "Conversation between President Chiang Kai-shek and Mr. Donald Nelson"; "Summary of Conversation Held in Chungking on September 12, 1944."

5. "Conversation between President Chiang Kai-shek and Mr. Donald Nelson," 4.

6. "Summary of Conversation Held in Chungking on September 12, 1944," 6–7.

7. Nelson, *Arsenal of Democracy*, 17.

8. Nelson, *Arsenal of Democracy*, xii.

9. Executive Order 9024, January 16, 1942, as quoted in James W. Fesler (War Production Board Historian), *Industrial Mobilization for War*, 208.

10. "Conversation between President Chiang Kai-shek and Mr. Donald Nelson," 8–10.

11. *China Handbook, 1937–1945*, 366–368; "Howard Coonley Report to Donald M. Nelson, April 30, 1945," reprinted in Gragg, "History of the American War Production Mission in China," 229–231.

12. "Organic Law Governing the Chinese War Production Board," in Gragg, "History of the American War Production Mission in China," 213–214.

13. Materials in Letters and Reports from War Production Board to Chairman Wen-Hao Wong.

14. "Howard Coonley Report to Donald M. Nelson, April 30, 1945," in Gragg, "History of the American War Production Mission in China," 238.

15. "Howard Coonley Report to Donald M. Nelson, April 30, 1945," in Gragg, "History of the American War Production Mission in China," 228–229.

16. "Howard Coonley Report to Donald M. Nelson, April 30, 1945," in Gragg, "History of the American War Production Mission in China," 230.

17. "Conversation between President Chiang Kai-shek and Mr. Donald Nelson," 2–3. It is unclear exactly whom Chiang was talking about. A number of the State Department's cultural relations experts (e.g., Dykstra, Lowdermilk, and Beltz) had actively sought to improve Chinese products and production. He may have been thinking of U.S. military advisers.

18. "Donald M. Nelson Report to the President, December 20, 1944," in Gragg, "History of the American War Production Mission in China," 9.

19. "LeRoy Whitney to E. A. Locke"; "H. LeRoy Whitney to Byron M. Bird." Whitney was technical adviser to the personal representative of the president.

20. "Louis L. Anderson to W. H. Graham."

21. See materials in HST, Papers of Edwin A. Locke Jr., Box 5.

22. "Members of the American Production Mission in China"; Robert M. Kerr, "American War Production Mission in China Personnel." The Walworth Company, founded in 1942, was an international company that focused on the manufacture of valves for a great variety of uses. McKinsey, Kearney & Co. was a consulting company. "H. LeRoy Whitney to Byron M. Bird."

23. "LeRoy Whitney to E. A. Locke."

24. Alice B. Ecke, "China: Huge Potential Market." A copy of this article can be found in FDR, AWPM, Box 4, indicating that the AWPM was collecting publications that discussed how China might become a market for U.S. goods.

25. "The 1945 Objectives of the Chinese War Production Board," in Gragg, "History of the American War Production Mission in China," 245.

26. "The 1945 Objectives of the Chinese War Production Board," in Gragg, "History of the American War Production Mission in China," 249.

27. "Walter G. Whitman to James Jacobson."

28. "Donald Nelson to Andrew Kearney."

29. "Howard Coonley Report to Donald M. Nelson, April 30, 1945," in Gragg, "History of the American War Production Mission in China," 229.

30. "The 1945 Objectives of the Chinese War Production Board," in Gragg, "History of the American War Production Mission in China," 250. U.S. military leadership in China attempted to reinforce this point about financing for CWPB procurement in a plan presented to Chiang Kaishek in mid-June 1945. See "Plan Presented to Generalissimo Chiang Kai-shek, June 16, 1945," in Gragg, "History of the American War Production Mission in China," 286.

31. "Weng Wenhao to Howard Coonley."

32. "Notes on a Conversation with Mr. Calvin Joyner."

33. "Lionel Booth to Edwin Locke."

34. "Henrik Oversen to Edwin Locke." Oversen is referring to Lu Zuofu's Min Sheng Shipbuilding Company.

35. "Ralph Strang to Lester Bosch."

36. "Howard Coonley to Weng Wen-hao," January 15, 1945. Interestingly, according to Wilma Fairbank, Reck, who did stay on for an extra year, argued in at least one memo in favor of China's adoption of the metric system, which would have been at odds

with the kinds of American industrial standards that Coonley and the AWPM were promoting, and that Weng and the CWPB were eager to adopt. See Fairbank, *America's Cultural Experiment*, 74.

37. "Howard Coonley to Whiting Willauer."

38. "Ralph Strang to Andrew Kearney."

39. "Roy M. Jacobs to Andrew Kearney."

40. As Ying Jia Tan has noted, "mobilization for war. . . created the urgency for greater standardization" in the power industry, a process that continued during the war, during which time, as Tan has shown, the NRC's Central Electrical Manufacturing Works in Kunming, as inland China's main producer of electrical components, became the "de facto national standards institute for the power sector." Tan, *Recharging China in War and Revolution*, 91, 103.

41. Ku, "National Bureau of Industrial Research," 1.

42. "Edward Waldschmidt to Herbert Graham."

43. Ku, "National Bureau of Industrial Research," 9.

44. Ministry of Agriculture and Forestry, "Report of the National Agricultural Research Bureau."

45. Shen, *Agricultural Resources*, 255–256.

46. Hume, "War Hitting American Tung-Oil Interests," 143; Chang, "China's Tung Oil and Its Future," reprinted in Chang, *In Search of China's Future*, 22.

47. "Minutes: 3rd Meeting of Liquid Fuel Advisory Committee."

48. "Edwin K. Smith to Patrick Hurley."

49. "Conversation between President Chiang Kai-shek and Mr. Donald Nelson," 3.

50. "Notes on a Conversation with Mr. Calvin Joyner."

51. Le Roy Whitney, "Transportation in China."

52. "Quantity and Model of Automobile in Free China"; "A General Sketch of Work Done on Auto-Parts Production." The CWPB actually estimated that only 40 percent of China's trucks were in service at any given moment.

53. "Weng Wenhao to Donald Nelson."

54. "Howard Coonley Report to Donald M. Nelson, April 30, 1945," reprinted in Gragg, "History of the American War Production Mission in China," 235–236.

55. Bosch, "Discussion with Dr. William F. Woo."

56. Gragg, "History of the American War Production Mission in China," 173–174.

57. Hu, "Some Isochronous Maps of China," 1.

58. Zhou et al., *Economical Atlas of Sichuan*, 66.

59. Teichman, *Journey to Turkistan*, 30.

60. Ho, "Reminiscences of Ho Lien," 338.

61. Zhou et al., *Economical Atlas of Sichuan*, 66.

62. Weller, *Caravan across China*, 283.

63. Lee, "Report of the Northwestern China Scientific Expedition," 3.

64. Needham, "Report on a Journey in the North West, Occupying Aug., Sept., Oct., Nov. 1943," in Needham, "Chinese Papers," 1:5.

65. Wan, "Work of National Highways Bureau," 52.

66. "National Highways Exhibition," 51.

67. "By Stage Transportation to Sinkiang," 58–59.

68. "Theodore Dykstra to Mrs. Dykstra," August 2, 1943, Dykstra Papers.

69. "Chinese Study American Methods."

70. "Translation of a Text of Telegram from Ambassador Wei Tao Min."

71. Gragg, "History of the American War Production Mission in China," 178.

72. See, for example, Dykstra letter to wife, September 7, 1943, and Dykstra letter to wife, February 26, 1944, Dykstra Papers; "Needham Complete Diaries," October 29, 1944.

73. "Agenda, 2nd Meeting of the Auto-Parts Advisory Committee."

74. "Ralph Strang to Lester Bosch."

75. Xie, "Our Petroleum Prospect."

76. Stallings, "Conference with Dr. H. Y. Chao."

77. Wang, "Alcohol Factories"; Stallings, "Alcohol in China."

78. "Production of Alcohol in Free China."

79. Conversation between Donald Nelson and Weng Wenhao as described in "List of China's Immediate Requirements."

80. See, for example, Wu, "Memorandum on China's Urgent Requirements," 2–4.

81. Stallings's misapprehension of the sorghum (gaoliang) situation was almost certainly informed by information he was getting from his Chinese colleagues. According to a report by K. H. Yang, a member of the CWPB's Liquid Fuel Advisory Committee on which Stallings also served, "Kaoliang is not the principal source of food in China, it being usually classified under the category of auxiliary food-stuff. Farmers generally plant kaoliang on hill side, or on slopes of their farmland, or along ditches of same which are not suitable for rice-planting." Yang, "Raw Materials for Alcohol" and "Minutes of the 1st Meeting of the Liquid Fuel Advisory Committee," 2, show that that committee's discussion reinforced this idea. "It is a wrong conception that increase of alcohol production greatly affects food supply as Kaoliang is rarely used for human food in this part of the country."

82. Roland, "Alcohol Report," 4. See also "Harold Roland to N. Meiklejohn" and "Walker, M. N. Memorandum to W. T. Stanton." Walker served as acting chief of the FEA's food division and argued to Stanton, also of the FEA, that given the number of acres planted in gaoliang and corn, it would not be possible to divert enough grain to produce thirty million more gallons of alcohol in 1945 without affecting food levels. "Andrew Kearney to Donald Nelson" indicates that like Roland, Kearney felt the figures did not support the conclusion that increased alcohol production would negatively affect food availability.

83. "Eugene Stallings to James Jacobson" and "Howard Coonley to Weng Wenhao," February 10, 1945.

84. "Draft Regulation of Liquid Fuel Control Commission."

85. Stallings, "Report on Meeting on Alcohol."

86. Paul H. Yen, manager of the Victory Alcohol Plant in Sichuan, mentioned many of these things in a November 28, 1944, report to Eugene Stallings that described in considerable detail the economic and policy challenges that alcohol producers faced. Stallings appears to have taken these comments to heart, and evidence shows that he also heard them from other sources. Yen, "Victory Alcohol Plant."

87. Stallings, "Alcohol in China."

88. Stallings, "Report on Meeting on Alcohol," 1–2.

89. Stallings, "Alcohol in China," 12.

90. Gragg, "History of the American War Production Mission in China," 12.

91. Gragg, "History of the American War Production Mission in China," 2.

92. "LeRoy Whitney to Hollis H. Arnold." Arnold had spent an extensive amount of time working as an electrical engineering consultant for the Nationalist government in the years right before the war. His résumé had crossed Whitney's desk in the summer of 1945, and Whitney clearly held on to it in hopes of building a private working relationship with him.

Chapter 5

1. Haven, "Report on Inspection of Iron and Steel Plants," 3.

2. Haven, "Report on Inspection of Iron and Steel Plants," 5.

3. Moyer, "Informal Summary No. 2." Moyer's collaborators were all connected to the Nationalist government, so his perception of what had been done in the occupied areas was shaped by their descriptions.

4. Locke, "Report to the President—December 18, 1945," in Gragg, "History of the American War Production Mission," 307.

5. See materials in Lend-Lease Training-General.

6. Willauer and Ray, as paraphrased in Foreign Economic Administration, "China Advisory Committee Meeting," 9.

7. Cheng, *Chinese Training Program, Final Report*, 1.

8. See Adas, *Dominance by Design*, 223–227.

9. See, for example, "Glen M. Ruby to C. D. Shiah"; "Notes from Mr. Pierce"; Locke, "Report to the President—December 18, 1945," appendix in Gragg, "History of the American War Production Mission," 304–316.

10. Allied leaders were convinced that international economic stability and a smoother system of international trade would foster peace. The Bretton Woods Conference of 1944, for example, promoted the idea of world peace through world trade. See Newcomer, "Bretton Woods and a Durable Peace," 37–42.

11. Hanson, "Blue Print for Exporting American Know-How," 5–6.

12. Hanson, "Blue Print for Exporting American Know-How," 7.

13. "Teaching United States Methods Builds Trade," 3.

14. "Primer for the Chinese."

15. "Letter from Major Chi-shan Chen to S. C. Wang," 1. Wang, of the China Supply Commission in Washington, DC, had previously been a professor of physics and quantum mechanics at Peking University and in the early 1940s served as the general manager of the State Central Machine Works.

16. "James Pierce to Shou-chin Wang," 20. Taub had written up the "Taub Industrialization Plan," a comprehensive plan for the postwar industrialization of China that the FEA, in particular, hoped might form a framework for China's postwar industrialization.

17. "Hilton Smith to Elliott Hanson."

18. Shire, "Abstract," 1.

19. Shire, "Abstract," 2.

20. Shire, "Abstract," 3.

21. *CIE News Bulletin* 4, no. 5 (October 10, 1945): 3.

22. For more detail on specific training programs and placements, see numerous files in the NRC archives, Academia Historica.

23. Eoyang, "Memo."

24. Jiang, "Final Report," 23.

25. Jiang, "Final Report," 23–24.

26. Liang, "Chinese Training Program," 1, 4.

27. Guo, "Final Report," 3.

28. Hwang, "Training Program for Chinese Technicians."

29. "Letter from Major Chi-shan Chen to S. C. Wang," 1.

30. Liu, "Chinese Training Program," 2.

31. Liu, "Chinese Training Program," 17–18.

32. Paine, *Wars for Asia*, 242–251.

33. Foreign Economic Administration, "China Advisory Committee Meeting," 6.

34. For example, when the Nationalist government requested specialists to do surveys, Gauss put his foot down, observing that there were plenty of Chinese who could do that work. "Gauss to Secretary of State," 700–701.

35. See Tan, *Recharging China in War and Revolution*, ch. 3.

36. In 1938, the Export-Import Bank had granted China a $25 million loan for the purchase of American manufactures it might need for defense. The loan was repaid with tung oil by the end of 1942.

37. "Edwin Locke to Weng Wenhao."

38. "Edwin Locke to Harry S. Truman."

39. Chiang, "Statement by President Chiang Kai-shek at First Meeting of Supreme Economic Council, Monday, November 26, 1945." Reproduced in Gragg, "History of the American War Production Mission," 300.

40. "Weng Wenhao to Locke," 1–2.

41. Janow, "American Technical and Administrative Personnel," 2–4.

42. United States Department of State, *China White Paper*, 226–227.

43. United States Department of State, *China White Paper*, 371–372.

44. United States Department of State, *China White Paper*, 399; U.S. Senate Committee on Foreign Relations, Report no. 1026, "Aid to China."

45. "Donald Nelson to T. V. Soong," 1–2.

46. Locke, "Report to the President—December 18, 1945," in Gragg, "History of the American War Production Mission," 309.

47. "Current Status of Rehabilitation Contracts," 1.

48. "Current Status of Rehabilitation Contracts." The Chemical Construction Company, which was the parent company of American Cyanamid Company, had previously supplied a technician for the AWPM.

49. "C. D. Shiah to Edwin Locke."

50. Lowery, "Notes on New York Visit," 2.

51. See documents in Contract for Technical Services with E. B. Badger and Sons Company.

52. Lowery, "Notes on New York Visit," 2.

53. See Cullather, "'Fuel for the Good Dragon,'" 6.

54. "Agreement," 3.

55. "C. Yun to W. A. Haven."

56. "Summary of Agreement."

57. "William A. Haven Curriculum Vita."

58. Haven, "Tentative Plan," 8. The report proposed that the items for which there was not yet a large market could be, for the time being, manufactured in some of the smaller mills the group had visited in Tianjin and Tangshan.

59. Haven, "Tentative Plan," 22.

60. See correspondence between Qian Changzhao and W. A. Haven in Arthur G. McKee and Company IC-4. In the fall of 1946, Qian took over as head of the NRC.

61. Haven, "Report on Inspection," 5.

62. See documents in Arthur G. McKee and Company IC-4.

63. "T. Y. Yen to Sandys Bao."

64. ErSelçuk, "Iron and Steel Industry," 21.

65. "Current Status of Rehabilitation Contracts," 5.

66. "Y. C. Young to O. R. Johnson." Note that Young, chairman of the Liquid Fuel Advisory Committee, was himself being sent to Taiwan in late August 1945 to help determine how best to take over Taiwan's Japanese-owned sugar industry. See "O. R. Johnson to James A. Jacobson."

67. See Sugar Industry Program Correspondences with Arthur G. Keller (Letters).

68. Rose, "Sugar Industry at Taiwan."

69. "C. H. Huang to C. D. Shiah."

70. "Channan Shen, General Manager, Taiwan Sugar Corporation (Foochow Rd, Shanghai), to NRC-NY."

71. "Cheng Paonan to R. G. A. Jackson."

72. See Taiwan Sugar-Bitting Incorporated, New York; Bitting, Incorporation (Sugar).

73. "T. S. Lee to L. F. Chen."

74. "C. D. Shiah to Gauss," 3–4.

75. "Interim Requirement for the Taiwan Sugar Industry," 1–2.

76. "Glen M. Ruby to Locke," 1.

77. "Glen M. Ruby to Locke," 2.

78. "Plan for the Development of Coal and Mineral Resources," 2.

79. "Notes from Mr. Pierce," 2.

80. "Notes from Mr. Pierce," 1.

81. "Notes from Mr. Pierce," 1–2.

82. Rose, "Sugar Industry at Taiwan," 3.

83. Rose, "Sugar Industry at Taiwan," 21.

84. Rose, "Sugar Refining Industry."

85. "Channan Shen, General Manager, Taiwan Sugar Corporation (Foochow Rd, Shanghai), to NRC-NY."

86. "Arthur Keller to C. D. Shiah." For comparison, Sam H. K. Shih, who participated in the agricultural extension training program at Iowa State University in 1945–

1946, indicated that the total cost per student of that course was about $6,300. See Shih, "Term Paper," 1.

87. "C. H. Wu Cable to C. D. Shiah and T. S. Lee."

88. "C. D. Shiah to Arthur G. Keller."

89. Chang, "Training Report."

90. "C. H. Huang to C. D. Shiah."

91. "Chen Che-pin to C. H. Shih."

92. "William L. C. Hui to Yuen Yin-kee." Yuen was an employee of the Peikang Sugar Factory in Taiwan. He arrived in the United States in October 1947 and sailed to China in August 1948. The other two trainees were Wang Hsun-shu (王舜緒), a chemical engineer, and Sun Chang-ling (孫常齡), studying industrial management. See also Reports of Yin-Kee Yuen S-406.

93. "The Secretary of State to the Embassy in China," 1274–1275.

94. "Mr. Leslie A. Wheeler to the Agricultural Attaché in China (Dawson)," 1277.

95. "Final Press Release of the China–United States Agricultural Mission," 1284–1285.

96. "Memorandum by the Agricultural Attaché in China (Dawson) to the Ambassador in China (Stuart)," 1289.

97. Shen, *Autobiography*, 178.

98. "Memorandum by T. H. Shen of the Chinese Ministry of Agriculture and Forestry," 1290–1292.

99. Shen, *Autobiography*, 180.

100. "Memorandum by the Chairman of the Chinese Supply Commission (Wang) to the Secretary of State."

101. See materials in Proposals to the Export-Import Bank 1947.

Epilogue

1. CNRRA, the Chinese counterpart institution created during the war to oversee the administration of UNRRA assistance and projects in China, was roundly criticized by American observers of the time as having been massively corrupt and having mismanaged the funds. See, for example, Kerr, *Formosa Betrayed*, chap. 8.

2. William Haven, "McKee Report on North China Trip," April 5, 1946; "Arthur G. McKee and Company IC-4."

3. See CIA, "Shih-Ching-Shan Iron and Steel Plant Shih-Ching-Shan, China," February 1, 1966. The fact that they were taking satellite photos and tracking activities there suggests that the plant was still important in 1966.

4. ErSelçuk, "The Iron and Steel Industry in China," 360, 366, 367. ErSelçuk says that plants in Chongqing were idle at the time the Chinese Communist Party took over the area and that those plants along with the ones at Shichingshan were rebuilt by the Communists using Russian plans and materials and that the labor for the Chongqing plants was all trained in Beijing between 1951 and 1953. Even if this is true, the plants, and particularly those in Chongqing, were themselves the products of Nationalist wartime industrial development efforts. ErSelçuk's sources for his report were all PRC media publications, so his information may not have been entirely accurate.

5. Zhan, Wang, and Deng, "The General View about the China Institute of Geography," 1768–1777.

6. Shen, *Unearthing the Nation*, 159.

7. Lowdermilk, "Soil, Forest, and Water," 396, 404–405.

8. Jansky et al., "Potato Production and Breeding in China," 62.

9. Lowdermilk, "Soil, Forest, and Water," 392–393.

10. "Jiang Shanxiang."

11. Ma, Shih-Ying, "Shih-Ying Ma Reports"; "Ma Shiying."

12. "Ching-Kwei Lee Reports"; "Li Jingkui."

13. Ouyang, "China's First Aircraft Manufacturing Plant and the Alumni of Zhejiang University."

14. Wu Daguan, "Wu Daguan's Oral Autobiography," especially 36–49.

15. Buchele, "Dr. Davidson's Program in China (January 1948 to December 1949)," 2.

16. GlobalSecurity.org, "816 Nuclear Military Plant."

17. "Chongqing Turns Its Dilapidated Factories."

18. See Zhao et al., "Research on Turning Tung Oil into Biodiesel," 180; Park et al., "Production and Characterization of Biodiesel from Tung Oil," 109–117; Shang et al., "Properties of Tung Oil Biodiesel and Its Blends with o# Diesel," 826–828; Shang et al., "Production of Tung Oil Biodiesel and Variation of Fuel Properties during Storage," 106–115; Yang et al., "Biodiesel Production from a Novel Raw Material Tung Oil," 406–410; Huo, "Production of Biodiesel from Tung Oil," 66–69; "A Method for the Production of Biodiesel Tung"; "Method for Recycling Tung Seed Press Cakes to Convert into Biodiesel"; "Method for Preparing Biodiesel from Tung Oil as Raw Material"; "Biodiesel Investments on the Increase in China." See also Global Subsidies Initiative, "Biofuels at What Cost?," 26.

19. New Zealand China Friendship Society, "Shandan Bailie School"; Li and Xuen, "Remote School with NZ Ties Trains Technicians."

20. Wang, "The Chinese Developmental State during the Cold War."

21. Fu Banghong has shown this reorientation toward state-planned science in the case of Academia Sinica in Fu, *Plan of Science and Planning of Science*.

22. Wang, "Chinese Developmental State," 182.

23. Fan, "A Study on the Guaranteed Sugar Price in Taiwan." The Council on Economic and Cultural Affairs, which funded Fan's master's-level training, was established by John D. Rockefeller III in 1953 as a nonprofit that promoted training and research in the social sciences in Asia.

24. Fan, "A Study," vii.

25. E. A. Rose, Inc., "The Sugar Industry at Taiwan," 8.

26. Vouchers of February 1952.

27. Loyalty Statements of Employees of the Taiwan Sugar Corporation—Xihu Factory.

28. General Information of Taiwan Power Company, 24.

29. See, for example, Lu and Huang, *Reminiscences of Mr. King Kai-Yung*. King had worked at the Yumen oil field and went to Taiwan with the Chinese Petroleum Corporation.

30. National Tsing Hua University Research Fellowship Fund and China Institute in America, *A Survey of Chinese Students in American Universities and Colleges*, 19–20;

Mutual Security Agency, "Monthly Report of the Mutual Security Agency to the Public Advisory Board," ii.

31. "C. Yun to W. S. Finlay Jr." See also Letters of White Engineering Works, Incorporated and Letters of the J. G. White Engineering Corp.

32. United States Economic Cooperation Administration, *Mission to China*, 37. It seems likely that the ECA's preference for working with a U.S. contractor stemmed from early postwar American skepticism about the Nationalist government's capacity (or willingness) to manage U.S. aid effectively.

33. "James H. Pierce to L. F. Chen."

34. Cullather, "Fuel for the Good Dragon," 6–7, 9, 18; United States Congress, Committee on Foreign Affairs, *Review of the Mutual Security Program Development and Implementation.*

35. United States Economic Cooperation Administration, *Mission to China*, 38.

36. A 1951 ECA report indicated that between late 1945 and the end of 1950, the Chinese government had spent $29.6 million on reconstruction efforts in Taiwan compared to $2.3 million from UNRRA and ECA. United States Economic Cooperation Administration, *Mission to China*, 45.

37. Tsai, "Innovation in Power Sources," 108–109.

38. Greene, *Origins of the Developmental State in Taiwan*, 80–87.

Bibliography

Archives

Academia Historica (AH)
Academia Sinica, Institute of Modern History
Cambridge University
Columbia University
Franklin D. Roosevelt Library (FDR)
Harvard University
Iowa State University
Nanjing Second Historical Archive
National Archives (UK)
National Archives (US)
Needham Research Institute
Truman Library (HST)
Wesleyan University

Journals and Series

Acta Brevia Sinensia
British Documents on Foreign Affairs: Reports and Papers from the Foreign Office Confidential Print (BDFA),
China at War
China Newsweek
Chinese Institute of Engineers News Bulletin
Dili xuebao (地理學報)
Far Eastern Survey
Foreign Relations of the United States (FRUS)
National Reconstruction Journal
Universal Engineering Digest

Books, Reports, Articles, Book Chapters, and Dissertations

"A Method for the Production of Biodiesel Tung." 2013. Accessed January 24, 2019. https://patents.google.com/patent/CN104560410B/en.

"A New System of Schools for Farmers under Experimentation." *Agriculture & Forestry Notes*. College of Agriculture and Forestry, University of Nanjing, Chengdu, No. 10, December 1940.

A Survey of Chinese Students in American Universities and Colleges in the Past One Hundred Years. New York: China Institute in America, 1954.

"A Working Program for the Improvement of Agricultural Education in China, Drafted by Dean Chang." *Agriculture & Forestry Notes*. College of Agriculture and Forestry, Jinling University, Chengdu, No. 10, December 1940.

"Academia Sinica's Northwest Scientific Investigation Team Cost, Budget, and Unit Work Plans, Etc." 中央研究院西北科學考察團經費概算及個組工作畫書等, Nanjing Second Archives, 393-139(1).

"The Acting Secretary of State to the Ambassador in China (Gauss)." October 3, 1942, FRUS, China, 1942.

Adas, Michael. *Dominance by Design: Technological Imperatives and America's Civilizing Mission*. Cambridge, MA: Harvard University Press, 2009.

"Agenda, 2nd Meeting of the Auto-Parts Advisory Committee." March 26, 1945. FDR, AWPM, Box 27, Folder: Transportation—Automotive and General.

"Agreement." March 14, 1945. Arthur G. McKee Company Contract Negotiation 阿瑟. 參琪公司與中國鋼鐵工業設計合同洽商交涉函件, AH 003-020400-0333.

Alley, Rewi. "CIC in the Northwest." *Gung Ho News*, January 1947.

———. *A Highway and an Old Chinese Doctor*. Christchurch: Caxton, 1973.

———. *Sandan: An Adventure in Creative Education*. Christchurch, New Zealand: Caxton, 1959.

———. "Shantan Bailie School, General Report, 1947." Shanghai: International Committee for Chinese Industrial Cooperatives, 1947.

"The Ambassador in China (Gauss) to the Secretary of State." March 27, 1942. FRUS, China, 1942.

"The Ambassador to China (Gauss) to the Secretary of State." October 7, 1943. FRUS, China-Diplomacy, 1943.

"The Ambassador in China (Gauss) to the Secretary of State." December 9, 1943. FRUS, China, 1943.

"The Ambassador in China (Gauss) to the Secretary of State." May 4, 1944. FRUS, China, 1944.

"The Ambassador in China (Gauss) to the Secretary of State." May 30, 1944. FRUS, China, 1944.

"The Ambassador in China (Gauss) to the Secretary of State." June 3, 1944. FRUS, China, 1944.

"The Ambassador in China (Gauss) to the Secretary of State." September 24, 1944. FRUS, China, 1944.

"Andrew Kearney to Donald Nelson." April 20, 1945. FDR, AWPM, Box 1, Liquid Fuels, Alcohol, General (Folder 1 of 2).

"Announcement of Opportunities for Practical Training." *CIE News Bulletin* 2, no. 4 (August 10, 1943).

"Announcing Fellowships for Graduate Chinese Students in the United States." Paul Meng Papers, Wesleyan University, Box 5, 1991.3.26.

Arkush, R. David. *Fei Xiaotong and Sociology in Revolutionary China*. Cambridge, MA: Harvard University Press, 1981.

Arthur G. McKee and Company IC-4 阿瑟麥琪公司 函件, AH 003-020100-0210.

"Arthur Keller to C. D. Shiah." January 21, 1947. Sugar Industry Program Correspondences with Arthur G. Keller (Letters) 糖工業與美國合作計畫交涉函件, AH 003-020 400-0225.

"Atcheson (Chargé in China) to Sec State." July 22, 1943. FRUS, China-Diplomacy, 1943.

Bailie, Victoria W. *Bailie's Activities in China*. Palo Alto, CA: Pacific Books, 1964.

Bao, C. M. 鮑覺民, and C. C. Chang 張景哲. "The Land Utilization of the Lo-Lung River Basin, Cheng-Kung District, Yunnan" 雲南省呈貢縣落老河區土地利用初步調查報告. *Dili xuebao* 11 (1944).

Barnett, Robert W. "China's Industrial Cooperatives on Trial." *Far Eastern Survey* 9, no. 5 (February 28, 1940): 51–56.

Bernstein, Richard. *China 1945: Mao's Revolution and America's Fateful Choice*. New York: Knopf, 2014.

Bian, Morris L. "Building State Structure: Guomindang Institutional Rationalization during the Sino-Japanese War, 1937–1945." *Modern China* 1 (2005): 35–71.

———. *The Making of the State Enterprise System in Modern China: The Dynamics of Institutional Change*. Cambridge, MA: Harvard University Press, 2005.

———. "Redefining the Chinese Revolution: The Transformation and Evolution of Guizhou's Regional State Enterprises, 1937–1957." *Modern China* 3 (2015): 313–350.

Bieler, Stacey. *"Patriots" or "Traitors"? A History of American-Educated Chinese Students*. Armonk, NY: M. E. Sharpe, 2004.

"Biodiesel Investments on the Increase in China." *Reliable Plant*. Accessed January 24, 2019. https://www.reliableplant.com/Read/9993/biodiesel-investments-on-increase-in -china.

Bitting, Incorporation (Sugar) 比亭公司—鹽糖案, AH 003-020200-0007.

Block, Jean Libman. "Potatoes for the Rice Bowls of China." *Saturday Evening Post*, October 20, 1945, 43.

Bosch, Lester. "Discussion with Dr. William F. Woo, Director of Southwest Region, CWPB, Kunming." July 3, 1945. FDR, AWPM, Box 27, Folder: Transportation—Automotive and General.

British Council. "Dr. Needham's Memorandum of 14th January, 1944, on Western Technical Experts in China." June 16, 1944. National Archives (UK), F3137/16/10.

———. "Industrial and Mining Exhibition of the National Resource Commission of the Ministry of Economic Affairs." *Acta Brevia Sinensia* 6 (April 1944).

Buchele, Wesley F. "Dr. Davidson's Program in China (January 1948 to December 1949)." October 21, 2005. Iowa State University Archives, Sherwood DeForest Papers, Box 9, Folder 4.

Buck, Peter. *American Science in Modern China, 1876–1936.* Cambridge: Cambridge University Press, 1980.

Bullock, Mary Brown. *The Oil Prince's Legacy: Rockefeller Philanthropy in China.* Washington, DC: Woodrow Wilson Center Press, 2011.

Burt, Sally K. *At the President's Pleasure: FDR's Leadership of Wartime Sino-US Relations.* Leiden, the Netherlands: Brill, 2015.

"By Stage Transportation to Sinkiang." *China at War* 12, no. 4 (April 1944).

"C. D. Shiah to Arthur G. Keller." Sugar Industry Program Correspondences with Arthur G. Keller (Letters) (糖工業與美國合作計畫交涉函件), AH 003-020400-0025.

"C. D. Shiah to Edwin Locke." June 12, 1946. HST, Edwin A. Locke Jr. Files, Box 9, Folder: National Resources Commission of China.

"C. D. Shiah to Gauss." October 23, 1947. Letters and Cables of Taiwan Sugar Corporation (臺灣糖業公司來往函件及電報), AH 003-020400-0281.

"C. H. Huang to C. D. Shiah." August 31, 1946. Cables of Taiwan Sugar Corporation (臺灣糖業公司電報), AH 003-020400-0159.

"C. H. Wu Cable to C. D. Shiah and T. S. Lee." March 14, 1947. Cables of Taiwan Sugar Corporation (臺灣糖業公司電報), AH 003-020400-0159.

"Chia Wei Chang to Chih Meng." April 14, 1942. National Archives, RG 114, Box No. 2, Records of the Assistant Chief, W. C. Lowdermilk, 1930–1947, China Trip—Clippings, Folder 4, Agricultural Aid for China, 1942.

"Chia Wei Chang to Lowdermilk." April 14, 1942. National Archives, RG 114, Box No. 2, Records of the Assistant Chief, W. C. Lowdermilk, 1930–1947, China Trip—Clippings, Folder 4, Agricultural Aid for China, 1942.

CIA. "Shih-Ching-Shan Iron and Steel Plant Shih-Ching-Shan, China." February 1, 1966. https://www.cia.gov/library/readingroom/docs/CIA-RDP78T05161A00060 0010052-6.pdf.

"C.I.E. Forum on Post-war Industrialization of China, Introduction." *National Reconstruction Journal* 3 (April 1943).

"C. Yun to W. A. Haven." October 9, 1945. Arthur G. McKee Company Contract Negotiation (阿瑟. 參琪公司與中國鋼鐵工業設計合同洽商交涉函件), AH 003-020400-0333.

Central Planning Bureau. "First Five-Year Program for China's Postwar Economic Development." N.p. January 1946.

Central Planning Bureau (中央設計局). *Five Year Plan for China's Economic Development* (Main Points) 物資建設五年計畫草案 (要提). N.p.: n.d.

Chang, Chia-chu. *In Search of China's Future.* San Francisco: I.B.C. Company, 1983.

"Chang Chia Wei to Chih Meng." April 14, 1942. National Archives, RG 114, Box No. 2, Records of the Assistant Chief, W. C. Lowdermilk, 1930–1947, China Trip—Clippings, Folder 4, Agricultural Aid for China, 1942.

Chang, Lie-Tien. "Training Report." May 11, 1946. Monthly Report of Trainee Chang, Lie-Tien 實習員張力田月報, AH 003-020600-2480.

"Channan Shen, General Manager, Taiwan Sugar Corporation (Foochow Rd, Shanghai), to NRC-NY." November 1, 1946. Letters of E. A. Rose, Incorporation 羅斯公司函件, AH 003-020400-0324.

"Check List for Use in a First Meeting of Leaders Interested in Postwar Planning." William Homer Stacy Papers, Iowa State University Archives, Box 14, Folder 5.

"Chen Che-pin to C. H. Shih." December 3, 1945. Che-Pin Chen Letters 2552 (陳濟平函件2552), AH 003,020600-1093.

Chen, Eugenia V. "Survey of Chinese Youth and Student Clubs in New York City—1945." MA thesis, University of Michigan, 1945.

"Chen, L. F. to Eaton." May 10, 1945. Letters of Professor Paul B. Eaton, AH 003-020600-1760.

"Chen, L. F. to Paul Eaton." May 11, 1945. Letters of Professor Paul B. Eaton, AH 003-020600-1760.

Chen, L. F. *Report to Annual Convention of the Technical Committee of the NRC.* December 27, 1944. 資源委員會駐美技術團年會紀錄 Annual Convention, 1944 TCG-6b, AH, 003-020100-0293.

Chen, Li-fu. *Chinese Education during the War (1937–1942).* Chongqing: Ministry of Education, 1943.

———. *The Storm Clouds Clear over China.* Stanford, CA: Hoover Institution Press, 1994.

Chen, Liang-fu. "Some Outstanding Problems of China's Post-war Industrialization." *National Reconstruction Journal* 3 (April 1943): 37–43.

Cheng, Hu Chi 程後琪. *Chinese Training Program, Final Report.* June 30, 1946. Trainee's Reports Management-Foreign Economic Administration (3) 在美受機械訓練報告, AH 003-020600-0482.

"Cheng Paonan to R. G. A. Jackson, UNRRA DC." May 1, 1946. Canadian Commercial Corporation Fertilizer 加拿大商業公司函件, AH 003-020100-0285.

Cheng, Yu-feng, and Cheng Yu-huang, eds. *Archives on the National Resources Commission Technicians' Training in the United States* 資源委員會技術人員赴美實習史料. Taipei: Academia Historica, 1988.

Ch'i, Hsi-sheng. *The Much Troubled Alliance: US-China Military Cooperation during the Pacific War, 1941–1945.* Singapore: World Scientific, 2016.

———. *Nationalist China at War.* Ann Arbor: University of Michigan Press, 1982.

Chiang, Kai-shek. *The Collected Wartime Messages of Generalissimo Chiang Kai-shek, 1937–1945.* Vol. 2. New York: John Day, 1946.

Chiang, Yung-chen. *Social Engineering and the Social Sciences in China, 1919–1949.* Cambridge: Cambridge University Press, 2001.

China Information Committee. *China after Four Years of War.* Chungking: China Publishing, 1941.

"China Institute in America, 60th Anniversary." Paul Meng Papers, Wesleyan University, Box 3, 1990.1.6.

China Travel Bureau 中國旅行設. "National Highways Exhibition, Special Issue" 全國公路展覽會特刊. 1944.

Chinese Institute of Engineers, Yunnan Branch. *Report.* FDR, Box 4, Folder "Chinese Institute of Engineers."

Chinese Ministry of Information. *China Handbook, 1937–1945.* New York: Macmillan, 1947.

"Chinese on the Job." *CIE News Bulletin* 2, no. 4 (August 10, 1943).

"Chinese Scholarship Winners Will Study at Iowa State." *Iowa State Daily Student*, October 26, 1945.

"Chinese Study American Methods." *Transit Lines*, August 1945. FDR, AWPM, Box 13, Folder: International Training Administration, Inc.—China.

"Chinese Studying in American Colleges Help Interpret Nation to Their Homeland." *Shanghai Evening Post and Mercury*, Amer. ed., New York, October 1, 1943.

"Ching-Kwei Lee Reports" 李慶逵報告, AH 003-020600-0477.

"Chongqing Turns Its Dilapidated Factories, Arsenals into Fashion and Arts Centers." *Global Times*, November 7, 2017. http://www.globaltimes.cn/content/1073984.shtml.

Clegg, Arthur. *The Birth of New China*. Watford, UK: Farleigh, 1943.

Clegg, Jenny. "Rural Cooperatives in China: Policy and Practice." *Journal of Small Business and Enterprise Development* 13, no. 2 (2006): 219–234.

Coble, Parks M. *China's War Reporters: The Legacy of Resistance against Japan*. Cambridge, MA: Harvard University Press, 2015.

Condliffe, J. B. "The Industrial Development of China." *National Reconstruction Journal* 3 (April 1943).

Contract for Technical Services with E. B. Badger and Sons Company (班吉桑斯公司檔案資料), AH 003-020400-0096.

"Conversation between President Chiang Kai-shek and Mr. Donald Nelson." September 19, 1944. FDR, AWPM, Box 36, Folder: China—Memoranda and Mission, Conferences, August–September 1944, Folder 2.

Coonley, Howard. "Memo: Answers to Questionnaire of Chinese Institute of Engineers, Yunnan Branch." January 26, 1945. FDR, Box 4, Folder: Chinese Institute of Engineers.

Cowan, C. D. *The Economic Development of China and Japan*. New York: Praeger, 1964.

Cullather, Nick. "'Fuel for the Good Dragon': United States and Industrial Policy in Taiwan, 1950–1965." *Diplomatic History* 20, no. 1 (Winter 1996): 1–25.

———. *The Hungry World: America's Cold War against Poverty in Asia*. Cambridge, MA: Harvard University Press, 2010.

"Current Status of Rehabilitation Contracts." April 13, 1946. HST, Edwin A. Locke Files, Box 11, Folder: NRCC Industrial Experts (National Resources Commission of China).

Davies, John Paton, Todd S. Purdum, and Bruce Cumings. *China Hand: An Autobiography*. Philadelphia: University of Pennsylvania Press, 2012.

"The Department of State to the British Embassy." FRUS: Diplomatic Papers, China, 1944, Volume VI. https://history.state.gov/historicaldocuments/frus1944v06/d707.

"Donald Nelson to Andrew Kearney." February 12, 1945. Letters and Reports from War Production Board to Chairman Wen-Hao Wong (戰時生產局到翁文灝函件及電文), AH 003-020100-0589.

"Donald Nelson to T. V. Soong." September 11, 1945. HST, Edwin A. Locke Jr. Files, Box 9, Folder: China Special.

Downey, Gary Lee, and Juan C. Lucena. "Knowledge and Professional Identity in Engineering: Code-Switching and the Metrics of Progress." *History and Technology* 20, no. 4 (December 2004): 393–420.

"Draft Regulation of Liquid Fuel Control Commission War Production Board." January 31, 1945. FDR, AWPM, Box 1, Folder: Liquid Fuels Control Commission.

Dykstra Papers, Harvard Yenching Library.

Eastman, Lloyd. *The Abortive Revolution: China under Nationalist Rule, 1927–1937.* Cambridge, MA: Harvard University Press, 1974.

———. *Seeds of Destruction: Nationalist China in War and Revolution, 1937–1949.* Stanford, CA: Stanford University Press, 1984.

Eaton, Paul E. "China's Industrial Training Program." *Far Eastern Survey* 13, no. 22 (November 1, 1944).

Ecke, Alice B. "China: Huge Potential Market for Textiles & Textile Machinery?" *Sales Management* (August 1944).

"Economic Milestones." *Free China Review* 16, no. 2 (February 1, 1966). http://taiwantoday.tw/ct.asp?xItem=156095&ctNode=2198&mp=9.

Edgerton, David. *The Shock of the Old.* London: Profile Books, 2006.

"Edward Waldschmidt to Herbert Graham." January 27, 1945. FDR, AWPM, Box 17, Folder: National Bureau of Industrial Research.

"Edwin Locke to Harry S. Truman." December 26, 1946. HST, Edwin A. Locke Jr. Files, Box 11, Folder: Pierce Management (Re: China).

"Edwin K. Smith to Patrick Hurley." April 9, 1945. FDR, AWPM, Box 16, Folder: Miscellaneous (Folder 4 of 4).

"Edwin Locke to Weng Wenhao." February 23, 1946. HST, Edwin A. Locke Jr. Files, Box 9, Folder: Weng Wen Hoa [*sic*].

"Employment Opportunities." *CIE News Bulletin* 3, no. 3 (June 10, 1944).

Eoyang, T. T. "Memo." December 31, 1948. Circular Letter—Close of Training Department 資源委員會裁撤訓練部公告, AH 003-020600-0174.

ErSelçuk, Muzaffer. "The Iron and Steel Industry in China." *Economic Geography* 32, no. 4 (October 1956).

Esherick, Joseph W. "Ten Theses on the Chinese Revolution." *Modern China* 1 (1995): 45–76.

———. *Lost Chance in China: The World War II Dispatches of John S. Service.* New York: Random House, 1974.

Esherick, Joseph W., and Matthew P. Combs, eds. *1943: China at the Crossroads.* Honolulu: University of Hawai'i Press, 2015.

"Eugene Stallings to James Jacobson." February 16, 1945. FDR, AWPM, Box 1, Folder: Liquid Fuels, Alcohol, General (Folder 2 of 2).

"Expeditions of Sinkiang." *Acta Brevia Sinensia* 6 (April 1944).

Fairbank, Wilma. *America's Cultural Experiment in China, 1942–1949.* Washington, DC: Department of State International Information and Cultural Series 108, 1976.

———. *Chinese Educational Needs and Programs of US-Located Agencies to Meet Them.* Paris: UNESCO, 1948.

Fan, Chwei-lin. "A Study on the Guaranteed Sugar Price in Taiwan." MS thesis, Montana State College, 1961.

Fan, Fa-ti. *British Naturalists in Qing China.* Cambridge, MA: Harvard University Press, 2004.

Farmer, Rhodes. *Shanghai Harvest: A Diary of Three Years in the China War.* London: Museum Press, 1945.

"Fazhan guofang kexue." Academia Sinica Institute of Modern History Archives, Ministry of Economics, 18-22-114.

Fei, Xiaotong, and Chang Zhiyi. *Earthbound China: A Study of Rural Economy in Yunnan.* Chicago: University of Chicago Press, 1945.

Fesler, James W. *Industrial Mobilization for War: History of the War Production Board and Predecessor Agencies, 1940–1945.* Vol. 1. Washington, DC: Historical Reports on War Administration, General Study No. 1, 1947.

"Final Press Release of the China–United States Agricultural Mission." November 16, 1946. FRUS, The Far East: China, 1946.

Fong, H. D. Fong, K. Y. Lin, and Tso-Fan Koh. "Problems of Economic Reconstruction in China," China Council Paper No. 2, Institute of Pacific Relations. In *The Modern Chinese Economy, Studies of Relief and Rehabilitation in China*, edited by Ramon Myers. New York: Garland, 1980.

Foreign Economic Administration. "China Advisory Committee Meeting." March 13, 1945. FDR, AWPM, Box 8.

"Foreword." *CIE Journal* 1, no. 1 (April 1943).

"Foreword." *Universal Engineering Digest* 1, no. 1 (April 1943).

Freyn, Hubert. *Chinese Education in the War.* Shanghai: Kelly and Walsh, 1940.

Fu, Banghong. *Plan of Science and Planning of Science in Republican China: An Investigation Centered on the Academia Sinica* 民国时期的科学计划与计划科学. Beijing: China Science and Technology Press, 2015.

———. "Science, Society and Planning: Joseph Needham's Report to Chiang Kaishek in 1946." *Cultures of Science* 3, no. 2 (2020): 97–117.

Fu, Runhua 傅潤華. *Kangzhan jianguo dahuashi* 抗戰建國大畫史. Shanghai: Zhongguo wenhua xintuo fuwushe, 1948.

"Gauss to Secretary of State." February 12, 1942. FRUS: Diplomatic Papers, 1942.

General Information of Taiwan Power Company 臺灣電力公司一般概況, AH 003-020100-0815.

"A General Sketch of Work Done on Auto-Parts Production Undertaken by Dept. of Manuf. WPB." FDR, AWPM, Box 27, Folder: Transportation—Automotive and General.

Geng, Xuan. "Serving China through Agricultural Science: American-Trained Chinese Scholars and 'Scientific Nationalism.' in Decentralized China (1911–1945)." PhD dissertation, University of Minnesota, 2015.

"Glen M. Ruby to C. D. Shiah." July 9, 1946. HST, Box 11, Folder: NRCC Industrial Experts (National Resources Commission of China).

"Glen M. Ruby to Locke." March 16, 1946. HST, Edwin A. Locke Jr. Files, Box 9, Folder: Hoover, Curtice & Ruby, Inc. (Re: Latin America and Petroleum in China).

GlobalSecurity.org. "816 Nuclear Military Plant, Baitao Township, Fuling District, Chongqing Municipality." Accessed November 24, 2019. https://www.globalsecurity.org/wmd/world/china/816-plant.htm.

Global Subsidies Initiative. "Biofuels at What Cost?" International Institute for Sustainability, November 2008.

Glover, Denise, Stevan Harrell, Charles F. McKhann, and Margaret Byrne Swain, eds. *Explorers & Scientists in China's Borderlands 1880–1950.* Seattle: University of Washington Press, 2011.

"Graff, E. F. Letter to W. H. Stacy." January 16, 1946. William Homer Stacy Papers, Iowa State University Archives, Box 14, Folder 5.

"Graff, E. F. Letter to W. H. Stacy et al." May 7, 1947. William Homer Stacy Papers, Iowa State University Archives, Box 14, Folder 5.

Gragg, Mabel. "History of the American War Production Mission in China." 1946. Truman Library, Papers of Harry S. Truman, Confidential File, Box 32.

Graham, Otis. *Toward a Planned Society: From Roosevelt to Nixon.* New York: Oxford University Press, 1976.

Green, O. M. *China's Struggle with the Dictators.* London: Hutchinson, 1941.

Greene, J. Megan. "Looking toward the Future: State Standardization and Professionalization of Science in Wartime China." In *Knowledge Acts in Modern China*, edited by Robert Culp, Eddy U, and Wen-hsin Yeh. Berkeley, CA: Institute of East Asian Studies, 2016.

———. *The Origins of the Developmental State in Taiwan.* Cambridge, MA: Harvard University Press, 2008.

Gross, Champ. "Aid to China's Agriculture." *Harvester World* 38, no. 2 (February 1947).

Gu Songfen 顾诵芬, ed. *Zhongguo feiji sheji de yi dai zong shi Xu Shunshou* 中国飞机设计的一代宗师 徐舜寿. Beijing: Hangkong gongye chubanshe, 2008.

Guo Bingyu 郭炳瑜. "Final Report." June 1946. Training Program for Chinese Technicians Tsun-Tsai Sun et al. 孫增在等八人訓練計畫, AH 003-020600-0515.

Guomin canzheng hui 國民參政會. *Guomin canzheng hui chuankang jianshe chatuan baogaoshu* 國民參政會川康建設察圖報告書. Taipei: Wenmei chubanshe, 1981.

"H. LeRoy Whitney to Byron M. Bird, Jeffrey Manufacturing Co. Columbus Ohio." May 30, 1945. HST, Papers of Edwin A. Locke Jr., Box 4, Folder: Job Application File D.

Hanson, Elliott S. "Blue Print for Exporting American Know-How." Address to the American Society of Mechanical Engineers, November 27, 1945. FDR, AWPM Box 13, Folder: International Training Administration, Inc.—China.

Hansen, Edwin L., Howard F. McColly, Archie A. Stone, and J. Brownlee Davidson. "Introducing Agricultural Engineering in China." *Agricultural and Biosystems Engineering Project Reports*, Paper 14 (1949).

Hao, K. S. "Southwest Gansu's Forest 甘肅西南之森林." *Dili xuebao* 9 (1942).

"Harold Roland to N. Meiklejohn of FEA in Kunming." May 9, 1945. FDR, AWPM, Box 1, Folder: Liquid Fuels, Alcohol, General (Folder 2 of 2).

Haven, William A. "McKee Report on North China Trip." April 5, 1946. Iron and Steel Industry Consulting Reports 鋼鐵事業顧問報告, AH 003-010700-0114.

———. "Report on Inspection of Iron and Steel Plants, Raw Materials and Transportation Facilities in Chungking District." Chungking, May 8, 1946. HST, Edwin A. Locke Jr. Files, Box 9, Folder: National Resources Commission of China.

———. "Tentative Plan for Development of Iron & Steel Industry in China." Report to National Resources Commission at Nanking by Arthur G. McKee & Company of Cleveland, Ohio, June 21, 1946. HST, Edwin A. Locke Jr. Files, Box 9, Folder: National Resources Commission of China.

Haw, Stephen G. "Tung Oil and Tong 桐 Trees." *Zeitschrift der Deutschen Morgenländischen Gesellschaft* (Journal of the German Oriental Society) 167, no. 1 (2017).

He, Fang 何芳. "Luo Jialun and the Northwest Expedition 罗家伦与西北建设考察团." MA thesis, Central China Normal University, Wuhan, 2012.

He, Simi, ed. *Documentary Collection on China and the Helping Hand of USA, 1937–1949* 抗戰時期美國援華史料. Taipei: Guoshiguan, 1994.

"Heard at the Inauguration Meeting." *CIE News Bulletin* 1, no. 1 (October 10, 1942).

"Henrik Oversen to Edwin Locke." September 25, 1945. FDR, AWPM, Box 30, Folder: Chinese Merit Awards—Personnel.

"Hilton Smith to Elliott Hanson." July 25, 1946. HST, Edwin A. Locke Jr. Files, Box 10, Folder: International Training Administration.

Ho, Lien. "The Reminiscences of Ho Lien (Franklin L. Ho)." Chinese Oral History Project, Columbia University.

Hogg, George. "Bailie Schools." February 26, 1944. INDUSCO, Ind., Records; Box 40, Folder: Hogg, George, articles/notes on Bailie Schools/typescript 1944, Rare Book and Manuscript Library, Columbia University.

———. "Letter to Ida Pruitt." June 21, 1942. INDUSCO, Ind., Records; Box 40, Folder: Hogg, George, articles/notes on Bailie Schools/typescript 1942, Rare Book and Manuscript Library, Columbia University.

———. "Letter to Ida Pruitt." July 23, 1944. INDUSCO, Ind., Records; Box 40, Folder: Hogg, George, articles/notes on Bailie Schools/typescript 1944, Rare Book and Manuscript Library, Columbia University.

———. "Plan for Lanchow Bailie School." June 1942. INDUSCO, Ind., Records; Box 40, Folder: Hogg, George, articles/notes on Bailie Schools/typescript 1942, Rare Book and Manuscript Library, Columbia University.

———. "A Report on the Shuangshihpu Bailie School." October 1942. INDUSCO, Ind., Records; Box 40, Folder: Hogg, George, articles/notes on Bailie Schools/typescript 1942, Rare Book and Manuscript Library, Columbia University.

Hostetler, Laura. *Qing Colonial Enterprise: Ethnography and Cartography in Early Modern China*. Chicago: University of Chicago Press, 2001.

Hou, Jiang 侯江. *Research on the Science Institute of West China* 中国西部科学院研究. Beijing: Zhongyang wenxian chubanshe, 2012.

Hou, T. P. "Nationalization of Chinese Heavy-Chemical Industries." *National Reconstruction Journal* 3 (April 1943).

"Hou Te-Bang." *Complete Dictionary of Scientific Biography*. New York: Scribner, 2008.

"Howard Coonley to Weng Wen-hao." January 15, 1945. Letters and Reports from War Production Board to Chairman Wen-Hao Wong 戰時生產局到翁文灝函件及電文, AH 003-020100-0589.

"Howard Coonley to Weng Wenhao." February 10, 1945. FDR, AWPM, Box 1, Folder: Liquid Fuels, Alcohol, General (Folder 2 of 2).

"Howard Coonley to Whiting Willauer." April 13, 1945. FDR, AWPM, Box 17, Folder: National Bureau of Industrial Research.

Howard, Joshua H. *Workers at War: Labor in China's Arsenals, 1937–1953*. Stanford, CA: Stanford University Press, 2004.

Hsiao, Theodore E. *The History of Modern Education in China*. Shanghai: Commercial Press, 1935.

Hsiung, James C., and Steven I. Levine, eds. *China's Bitter Victory: The War with Japan, 1937–1945*. Armonk, NY: M. E. Sharpe, 1992.

Hsu, Madeline Yuan-yin. *The Good Immigrants: How the Yellow Peril Became the Model Minority*. Princeton, NJ: Princeton University Press, 2015.

Hu, Danian. "American Influence on Chinese Physics Study in the Early Twentieth Century." *Physics in Perspective* 17 (2016): 268–297.

———. "The Reception of Relativity in China." *Isis* 98 (2007): 539–557.

Hu, Huan-Yong 胡焕庸. "Some Isochronous Maps of China" 国内交通與等時線圖. *Dili xuebao* 3, no. 4 (1936).

Hu, Powei 胡伯威. *"Childhood" Republican Period* "兒時" 民國. Guangxi shifan daxue. Accessed July 13, 2011. http://mjlsh.usc.cuhk.edu.hk/book.aspx?cid=2&tid=1&pid=705.

Huang, H. T. "Peregrinations with Joseph Needham in China, 1943–1944." In *Explorations in the History of Science and Technology in China*, edited by Li Guohao, Zhang Mengwen, and Cao Tianqin. Shanghai: Shanghai Chinese Classics Publishing House, 1982.

Hume, Joy. "War Hitting American Tung-Oil Interests." *Far Eastern Survey* 8, no. 12 (June 7, 1939).

Huo, W. Z. "Production of Biodiesel from Tung Oil." *China Oils and Fats* 39, no. 10 (2014): 66–69.

Hwang Yi 黃翼. "Training Program for Chinese Technicians, Monthly Report." January 1946. Trainee's Reports Management—Foreign Economic Administration (3) 在美受機械訓練報告, AH 003-020600-0483.

"Interim Requirement for the Taiwan Sugar Industry." December 1947. Taiwan Sugar Industry 臺灣糖業工廠, AH 003-020200-0034.

International Committee for Chinese Industrial Cooperatives. "Shantan Bailie School Training School for the Chinese Industrial Co-operatives." 1948. In Victoria W. Bailie, *Bailie's Activities in China*. Palo Alto: Pacific Books, 1964.

International Harvester. "For Release Immediately." January 5, 1945. http://content.wisconsinhistory.org/cdm/compoundobject/collection/ihc/id/48952/show/48888/rec/16.

International Harvester Company. "A Report on Agriculture and Agricultural Engineering in China." Chicago: International Harvester, 1949.

International Harvester News Releases: 1947. http://content.wisconsinhistory.org/cdm/compoundobject/collection/ihc/id/49350/show/49348/rec/15.

"Interview with Dr. Wong Wen Hao." Monday, March 28, 1945. HST, John D. Sumner Papers, Box 1, Folder: China Files, Chinese Opinions on Post-war Development, 1932–1944.

Israel, John. *Lianda: A Chinese University in War and Revolution*. Stanford, CA: Stanford University Press, 1998.

"James Pierce to Shou-chin Wang." June 13, 1945. Letters of James H. Pierce 詹姆士。皮爾斯函件, AH 003-020400-0156.

Jansky, S. H., L. P. Jin, K. Y. Xie, C. H. Xie, and D. M. Spooner. "Potato Production and Breeding in China." *Potato Research* 52 (2009): 57–65.

Janow, S. J. "American Technical and Administrative Personnel for the Chinese." March 28, 1946. HST, Edwin A. Locke Jr. Files, Box 7, Folder: China General, 1946.

Jen, Mei-Ngo, S. P. Chen, L. P. Yang, Y. F. Shih, and S. C. Chiao. "Land Utilization in Tsunyi District, Kweichow." *Dili xuebao* 13 (1946).

"Jiang Shanxiang." Accessed November 13, 2019. https://baike.baidu.com/item /%E6%B1%9F%E5%96%84%E8%A5%84.

Jiang Shanxiang 江善襄. "Final Report." June 28, 1946. Academia Historica, NRC Files, 003-020600-0515, 孫增在等八人訓練計畫, Training Program for Chinese Technicians Tsun-Tsai Sun et al.

Kerr, George. *Formosa Betrayed*. Boston: Houghton Mifflin, 1965.

Kerr, Robert M. "American War Production Mission in China Personnel: November 1944 through June 30, 1945." June 30, 1945. FDR, AWPM, Box 18, Folder: Personnel Reports, Lists and Miscellaneous (Folder 1 of 2).

Koen, Ross Y. *The China Lobby in American Politics*. New York: Octagon Books, 1974.

Kinzley, Judd C. "Crisis and the Development of China's Southwestern Periphery: The Transformation of Panzhihua, 1936–1969." *Modern China* 5 (2012): 559–584.

———. "Staking Claims to China's Borderlands: Oil, Ores and Statebuilding in Xinjiang Province, 1893–1964." PhD diss., University of California, San Diego, 2012.

Kirby, William C. "Continuity and Change in Modern China: Economic Planning on the Mainland and on Taiwan, 1943–1958." *Australian Journal of Chinese Affairs* 24 (1990): 121–141.

———. "Engineering China: Birth of the Developmental State, 1928–1937." In *Becoming Chinese, Passages to Modernity and Beyond*, edited by Yeh Wen-hsin. Berkeley: University of California Press, 2000.

Ku, Y. T. "National Bureau of Industrial Research, Ministry of Economic Affairs, a Brief Sketch of Its Work." FDR, AWPM, Box 17, National Bureau of Industrial Research Folder.

Kwan, Man Bun. *Patriot's Game: Yongli Chemical Industries, 1917–1953*. Leiden, the Netherlands: Brill, 2016.

Kwoh, Edwin Sih-Ung. "Chinese Students in American Universities: A Report of a Type C Project." EDD thesis, Columbia University, 1946.

Lam, Tong. *A Passion for Facts: Social Surveys and the Construction of the Chinese Nation State, 1900–1949*. Berkeley: University of California Press, 2011.

Lary, Diana. *The Chinese People at War: Human Suffering and Social Transformation, 1937–1945*. Cambridge: Cambridge University Press, 2010.

Lawson, Joseph. "Unsettled Lands: Labour and Land Cultivation in Western China during the War of Resistance (1937–1945)." *Modern Asian Studies* 5 (2015): 1442–1484.

Lean, Eugenia. *Vernacular Industrialism in China*. New York: Columbia University Press, 2020.

Lecuyer, Christophe. "The Making of a Science Based Technological University: Karl Compton, James Killian, and the Reform of MIT, 1930–1957." *Historical Studies in the Physical and Biological Sciences* 23, no. 1 (1992): 153–180.

Lee, J. S. *The Geology of China*. London: Thomas Murby, 1939.

Lee, Seung-Joon. "The Patriot's Scientific Diet: Nutrition Science and Dietary Reform Campaigns in China, 1910s–1950s." *Modern Asian Studies* 6 (2015): 1808–1839.

Lee, Shu-Tang. "Report of the Northwestern China Scientific Expedition 西北科学考察紀略." *Dili xuebao* 9 (1942): 1–30.

Lei, Sean Hsiang-lin. *Neither Donkey nor Horse: Medicine in the Struggle over China's Modernity*. Chicago: University of Chicago Press, 2014.

Lend-Lease Training-General 相借法案訓練案— 一般函件, AH 003-020600-0075.

Lepore, Jill. "Our Own Devices." *New Yorker*, May 12, 2008.

"LeRoy Whitney to E. A. Locke." January 26, 1945. HST, Edwin A. Locke Jr. Files, Box 4, Folder: Job Application File B.

"LeRoy Whitney to Hollis H. Arnold." August 28, 1945. HST, Edwin A. Locke Jr. Files, Box 4, Folder: Job Application File A.

"Letter from Chen Lifu to Jiang Tingfu on the List of American Experts Needed by the Ministry of Education" (*Chen Lifu han Jiang Tingfu Jiaoyubu suo xu mei zhuanjia mingdan*). In *Documentary Collection on China and the Helping Hand of USA, 1937–1949* (*Kangzhan shiqi meiguo yuan hua shiliao*), edited by He Simi. Taipei: Guoshiguan, 1994.

"Letter from Major Chi-shan Chen to S. C. Wang." September 18, 1945. S. C. Wang (王守兢函件), AH 003-020600-0253.

"Letter from Patrick J. Hurley to T. V. Soong, Chungking," December 26, 1944. FDR, AWPM, Box 25.

"Letter from Y. T. Ku to William H. Dennis." Nanjing Second Historical Archive, 448–543.

"Letter to Dr. Ku and Co-workers from Wm. Han-chu Lee et al." January 1945. Nanjing Second Historical Archive, NBIR 448–540.

Letters and Reports from War Production Board to Chairman Wen-Hao Wong 戰時生產局到翁文灝函件及電文, AH 003-020100-0589.

Letters of the J. G. White Engineering Corp. J. G. 懷特工程公司函件, AH 003-020400-0298.

Letters of White Engineering Works, Incorporated 懷特機械工程公司函件, AH 003-020600-2086.

Li, Danke. *Echoes of Chongqing: Women in Wartime China*. Urbana: University of Illinois Press, 2010.

Li, Hanhun. "The Reminiscences of Li Hanhun, as Told to Minta C. Wang, April 15, 1961, to February 3, 1962." Chinese Oral History Project, East Asian Institute, Columbia University.

"Li Jingkui." Accessed November 13, 2019. https://baike.baidu.com/item/%E6%9D%8E%E5%BA%86%E9%80%B5.

Li, Ming. "Post-war Industrialization Problems in China." *National Reconstruction Journal* 3 (April 1943): 30–31.

Li, Yang, and Xuen Chaohua. "Remote School with NZ Ties Trains Technicians." *China Daily*, November 17, 2015. ChinaDaily.com, http://www.chinadaily.com.cn/china/2015-11/17/content_22468625.htm.

Li, Yao Tzu. *Freedom & Enlightenment: My Life as an Educator/Inventor in China and the United States*. Lanham, MD: Lexington Press, 2003.

Liang, Sian 梁燊. "Chinese Training Program, Final Report." June 1946. Academia Historica, NRC Files, 003-020600-0481 在美受機械訓練報告 Trainee's Report Management—Foreign Economic Administration (1).

Lin, Hsiao-ting. "Nationalists, Muslim Warlords, and the 'Great Northwestern Development' in Pre-Communist China." *China & Eurasia Forum Quarterly* 1 (2007): 115–135.

Lindee, M. Susan. *Rational Fog: Science and Technology in Modern War*. Cambridge, MA: Harvard University Press, 2020.

"Linking the Northwest and the Southwest." *China at War* 12, no. 1 (January 1944).

"Lionel Booth to Edwin Locke." October 1, 1945. FDR, AWPM, Box 30, Folder: Chinese Merit Awards—Personnel.

"List of China's Immediate Requirements." September 1944. FDR, AWPM, Box 35, Memoranda on Mission Conferences (Folder 1 of 3).

Liu, Jiaping 劉嘉平. "Exploring the Secrets of Aircraft Engine Factory—Visiting Mr. Wang Wenhuan to Explore the Wonders That Occurred in Crow Cave 大定廠探密— 訪王文煥先生談烏鴉洞的奇蹟." *Zhonghua keji shixuehui xuekan* 16 (December 2011).

Liu Shoudao 劉守道. "Chinese Training Program, Final Report." June 30, 1946. Trainee's Reports Management—Foreign Economic Administration (3) 在美受機械訓練報告, AH 003-020600-0483.

"Lockhart to Orchard, Memorandum on Points of Possible Discussion with Mr. T. V. Soong." May 3, 1945. National Archives, 893.50/5345.

"Louis L. Anderson to W. H. Graham." November 17, 1944. HST, Edwin A. Locke Jr. Files, Box 4, Folder: Job Application File A.

Lowdermilk, Walter C. "Preliminary Report to the Executive Yuan, Government of China, On Findings of a Survey of a Portion of the Northwest for a Program of Soil, Water and Forest Conservation." FDR, AWPM, Box 1, Agriculture and Forestry, Folder 2.

———. "Soil, Forest, and Water Conservation and Reclamation in China, Israel, Africa, and the United States." An interview conducted by Malca Chall, Berkeley, 1969.

———. "Statement for News Notes for Chinese Students." January 25, 1944. National Archives, RG 114, Box No. 2, Records of the Assistant Chief, W. C. Lowdermilk, 1930–1947, China Trip—Clippings, Folder 6, Newsnotes for Chinese Students, 1943–1945.

"Lowdermilk, Walter C., to Chia Wei Chang." May 15, 1942. National Archives, RG 114, Box No. 2, Records of the Assistant Chief, W. C. Lowdermilk, 1930–1947, China Trip—Clippings, Folder 4, Agricultural Aid for China, 1942.

"Lowdermilk, Walter C., to Chih Meng." May 12, 1942. National Archives, RG 114, Box No. 2, Records of the Assistant Chief, W. C. Lowdermilk, 1930–1947, China Trip—Clippings, Folder 4, Agricultural Aid for China, 1942.

"Lowdermilk, W. C., to D. Y. Lin." January 31, 1942. National Archives, RG 114, Box No. 2, Records of the Assistant Chief, W. C. Lowdermilk, 1930–1947, China Trip—Clippings, Folder, China Material 1940–1942.

"Lowdermilk, W. C., to H. G. Calkins." Regional Conservator, Soil Conservations Service, Albuquerque, May 15, 1942. National Archives, RG 114, Box No. 2, Records of the Assistant Chief, W. C. Lowdermilk, 1930–1947, China Trip—Clippings, Folder 4, Agricultural Aid for China, 1942.

Lowery, Arthur. "Notes on New York Visit of 20 June 1946." June 21, 1946. HST, Edwin A. Locke Jr. Files, Box 9, Folder: National Resources Commission of China.

Loyalty Statements of Employees of the Taiwan Sugar Corporation—Xihu Factory 臺灣糖業公司職員失忠切結書— 溪湖廠, AH 003-010102-2624.

Lu, Pao-ch'ien, and Huang Ming-ming, eds. *The Reminiscences of Mr. King Kai-Yung* 金開英先生訪問紀錄. Taipei: Institute of Modern History Oral History Series, Academia Sinica, 1991.

Luo, Jialun. *Report on the Northwest Reconstruction Investigative Expedition* 西北建設考察團報告. Taipei: Guoshiguan, 1969.

"Ma Baozhi: China's Ministry of Agriculture Scholars of the War of Resistance Period" 馬保之: 抗戰時"中國農部"掌門人. Accessed August 31, 2016. http://www.njzy99.com/show.asp?news_id=3138.

Ma Baozhi and the Liuzhou Animal Husbandry School 馬保之与柳州牧校. Accessed August 31, 2016. http://blog.sina.com.cn/s/blog_63ad4fd70100mw21.html.

Ma, Shih-Ying. "Shih-Ying Ma Reports," 馬世英報告, AH 003-020600-0478.

"Ma Shiying." Accessed November 13, 2019. https://baike.baidu.com/item/%E9%A9%AC%E4%B8%96%E8%8B%B1/5586947.

MacKinnon, Stephen R. *Wuhan, 1938: War, Refugees, and the Making of Modern China*. Berkeley: University of California Press, 2008.

MacKinnon, Stephen R., Diana Lary, and Ezra F. Vogel, eds. *China at War: Regions of China, 1937–1945*. Stanford, CA: Stanford University Press, 2007.

MacNair, Harley Farnsworth, ed. *Voices from Unoccupied China*. Chicago: University of Chicago Press, 1944.

Marx, Leo. "Technology: The Emergence of a Hazardous Concept." *Technology and Culture* 51, no. 3 (2010): 561–577.

"May 22, 1945 Agreement on Services between ITA and the NRC Technical Office." AH 003-020600-1785.

May, Ernest R. *The Truman Administration and China, 1945–1949*. Philadelphia: Lippincott, 1975.

Mei, Yi-chi, ed. *A Survey of Chinese Students in American Universities and Colleges in the Past One Hundred Years*. New York: China Institute in America, 1954.

"Members of the American Production Mission in China." April 1946. FDR, AWPM, Box 18, Folder: Personnel Reports, Lists and Miscellaneous (Folder 1 of 2).

"Memorandum by Mr. Haldore Hanson of the Division of Cultural Relations." July 15, 1942. FRUS, China, 1942.

"Memorandum by T. H. Shen of the Chinese Ministry of Agriculture and Forestry." FRUS, The Far East: China, 1946.

"Memorandum by the Agricultural Attaché in China (Dawson) to the Ambassador in China (Stuart)." November 25, 1946. FRUS, The Far East: China, 1946.

"Memorandum by the Chairman of the Chinese Supply Commission (Wang) to the Secretary of State." February 17, 1947. FRUS, The Far East: China, 1947.

"Memorandum of Conversation, by the Ambassador in China (Gauss)." April 12, 1944. FRUS, China, 1944.

"Memorandum of Conversation, by the Special Assistant in the Division of Science, Education and Art (Peck)." May 17, 1944. FRUS, China, 1944.

"Memorandum Prepared in the Department of State." July 10, 1942. FRUS, China, 1942.

Meng, Chih. "China Institute: A Chinese-American Partnership." Paul Meng Papers, Wesleyan University, Box 6, 1991.3.42.

———. "Chinese-American Educational and Cultural Relations." 1944. Paul Meng Papers, Wesleyan University, Box 5, 1991.3.26.

———. *Chinese American Understanding: A Sixty-Year Search.* New York: China Institute in America, 1981.

———. "Committee on Wartime Planning for Chinese Students in the United States, Report of the Executive Director for the Period from September 1, 1944 to February 28, 1945." Paul Meng Papers, Wesleyan University, Box 5, 1991.3.32.

———. "Foreword." *National Reconstruction* 4 (February 1944).

———. "Foreword." *National Reconstruction* 5 (July 1944).

———. "National Reconstruction Forums." 1943. Paul Meng Papers, Wesleyan University, Box 5, 1991.3.26.

———. "Training of Chinese Technical Personnel in the United States." *National Reconstruction* 3 (April 1943).

"Method for Preparing Biodiesel from Tung Oil as Raw Material." 2016. https://patents.google.com/patent/CN106433997A/en.

"Method for Recycling Tung Seed Press Cakes to Convert into Biodiesel." 2014. https://patents.google.com/patent/CN105886127A/en.

Ministry of Agriculture and Forestry. "Report of the National Agricultural Research Bureau, 1932–44." HST, John D. Sumner Papers, Box 1, Folder: China Files: Agriculture Report of the National Agricultural Research Bureau, 1932–1944.

Ministry of Economic Affairs. *Guide to Industrial & Mining Exhibition of National Resources Commission.* Chungking: Ministry of Economic Affairs, National Government of China, 1944.

———. "Report of the Works of the Ministry of Economic Affairs." August 1944. FDR, Box 10, Folder: Industry, General Reports (Folder 3 of 3).

Ministry of Education. *University Course Goals* 大學科目標. Chongqing: Zhengzhong shuju, 1946.

"Minutes of the 1st Meeting of the Liquid Fuel Advisory Committee." December 12, 1944. FDR, AWPM, Folder: Liquid Fuels, Alcohol—Meetings and Conferences (Folder 2 of 2).

"Minutes: 3rd Meeting of Liquid Fuel Advisory Committee, WPB." January 25, 1945. FDR, AWPM, Box 1, Liquid Fuels, Alcohol, General (Folder 1 of 2).

"More Chinese Students Arrive." *Iowa State Daily Student* 74, no. 1 (September 22, 1945).

Moyer, Raymond. "Informal Summary No. 2: Covering Activities of the China–United States Agricultural Mission, July 15 to August 26, 1946." August 20, 1946. HST, Edwin A. Locke Jr. Files, Box 7, Folder: China General, 1946.

"Mr. Eden to Sir H. Seymour." July 27, 1943. In *British Documents on Foreign Affairs: Reports and Papers from the Foreign Office Confidential Print,* edited by Paul Preston and Michael Partridge, Part II, Series E, Volume 7. Arlington, VA: University Publications of America, 1997.

"Mr. K. N. Chang's Speech." January 11, 1944. *CIE News Bulletin* 3, no. 1 (February 10, 1944).

"Mr. Leslie A. Wheeler to the Agricultural Attaché in China (Dawson), at Shanghai." March 27, 1946. FRUS, The Far East: China, 1946.

Mueggler, Erik. *The Paper Road: Archive and Experience in the Exploration of West China and Tibet*. Berkeley: University of California Press, 2011.

Muscolino, Micah. *The Ecology of War in China: Henan Province, the Yellow River and Beyond, 1938–1950*. Cambridge: Cambridge University Press, 2015.

———. "Refugees, Land Reclamation, and Militarized Landscapes in Wartime China: Huanglongshan, Shaanxi, 1937–45." *Journal of Asian Studies* 69, no. 2 (May 2010): 453–478.

Mutual Security Agency. "Monthly Report of the Mutual Security Agency to the Public Advisory Board." August 31, 1952.

Nanking Academic, Related to Faculty and Staff. RG011-198-3399, Yale Divinity School Library. Accessed September 14, 2020. http://divinity-adhoc.library.yale.edu/UnitedBoard/University_of_Nanking/Box%20198/RG011-198-3399.pdf.

"National Highways Exhibition." *China at War* 12, no. 5 (May 1944).

National Bureau of Industrial Research. "NBIR Work Plan 中工所工作計畫." Nanjing Second Archives, 448-2313.

National Resources Commission. *Guide to Industrial & Mining Exhibition*. Chungking: Ministry of Economic Affairs, 1944.

National Tsing Hua University Research Fellowship Fund and China Institute in America. *A Survey of Chinese Students in American Universities and Colleges in the Past One Hundred Years*. New York: China Institute in America, 1954.

Needham, Joseph. "Chinese Papers, 1942–1946." 2 vols. NRI.

———. "Chungking Industrial and Mining Exhibition." *Nature* 153, no. 3892 (June 3, 1944).

———. "Journal of North-West Tour, August–December 1943." NRI.

———. "Letter to Sir Ronald." October 23, 1947. National Archives (UK), BW 23/4, Bailie Schools.

———. "Memorandum on Western Technical Experts in China." January 14, 1944. National Archives (UK), F3137/16/10.

———. Needham Archives, University Library, Cambridge.

———. "Needham Complete Diaries." NRI.

———. "Science and Technology in the North-West of China." *Nature* 153, no. 3878 (February 26, 1944).

———. "Science in Chungking." *Nature* 152, no. 3846 (July 17, 1943).

———. "Science in South-West China." *Nature* 152, no. 3845 (July 10, 1943).

Needham Research Institute Collection of Wartime Photos. http://www.nri.org.uk/JN_wartime_photos/nw.htm.

Nelson, Donald M. *Arsenal of Democracy: The Story of American War Production*. New York: Harcourt, Brace, 1946.

New Zealand China Friendship Society. "Shandan Bailie School." Accessed November 24, 2019. http://nzchinasociety.org.nz/shandan-bailie-school/.

Newcomer, Mabel. "Bretton Woods and a Durable Peace." *Annals of the American Academy of Political and Social Science* 240 (July 1945): 37–42.

Ni, Ting. *The Cultural Experiences of Chinese Students Who Studied in the United States during the 1930s–1940s.* Lewiston, UK: Edwin Mellen, 2002.

Northwest Branch of the National Geological Survey 中央地質調察所西北分所. *Overview of the Northwest Branch of the National Geological Survey* 中央地質調察所西北分所概. Nanjing: Zhongyang dizhi diaochasuo, 1948.

"Notes from Mr. Pierce." June 28, 1946. HST, Edwin A. Locke Jr. Files, Box 11, Folder: Pierce Management (Re: China).

"Notes on a Conversation with Mr. Calvin Joyner, F.E.A. Representative in Chungking." September 11, 1944. FDR, AWPM, Box 27, Transportation—Automotive and General.

"Notes on Conference Held with Dr. T. V. Soong." December 21, 1944. FDR, AWPM, Box 25, Folder: T. V. Soong.

"Notes on March 13, 1942 Meeting Called by L. A. Wheeler on Agricultural Assistance to China." National Archives, RG 114, Box No. 2, Records of the Assistant Chief, W. C. Lowdermilk, 1930–1947, China Trip—Clippings, Folder: China Material 1940–1942.

Ogle, Vanessa. "Whose Time Is It? The Pluralization of Time and the Global Condition, 1870s–1940s." *American Historical Review* 118, no. 5 (December 2013): 1376–1402.

"O. R. Johnson to James A. Jacobson." August 27, 1945. FDR, Box 10, Folder: Formosa.

"Organizational Report." *National Reconstruction Journal* 1 (August 1942).

Orleans, Leo A. *Professional Manpower and Education in Communist China.* Washington, DC: National Science Foundation, 1961.

Orleans, Leo A., ed. *Science in Contemporary China.* Stanford, CA: Stanford University Press, 1980.

Ouyang Changyu. "China's First Aircraft Manufacturing Plant and the Alumni of Zhejiang University 中国第一个航空发动机制造厂与浙江大校友." Accessed November 15, 2019. http://zuaa.zju.edu.cn/publication/article?id=74.

Paine, S. C. M. *The Wars for Asia, 1911–1949.* Cambridge: Cambridge University Press, 2012.

Pan, S. C. "A Plan Proposed for Improving Alcohol Industry in China." FDR, American War Production Mission in China, Box 2, Folder: Liquid Fuels, Alcohol "A Plan Proposed for Improving Alcohol Industry in China."

Park, J. Y., D. K. Kim, Z. M. Wang, P. Lu, S. C. Park, and J. S. Lee. "Production and Characterization of Biodiesel from Tung Oil." *Applied Biochemistry and Biotechnology* 148, no. 1–3 (March 2008): 109–117.

Payne, Robert. *China Awake.* London: Heinemann, 1947.

———. *Chungking Diary.* London: Heinemann, 1945.

Peck, Graham. *Through China's Walls.* London: Collins, 1941.

———. *Two Kinds of Time.* Cambridge: Houghton Mifflin, 1950.

Peng, Wen-ho. "A Program for Sheep Improvement and Wool Research Work in China." Academia Sinica, Institute of Modern History Archive, File: 20-25-53-5.

Peraino, Kevin. *A Force So Swift: Mao, Truman, and the Birth of Modern China, 1949.* New York: Crown, 2017.

Perleberg, Max. *Who's Who in Modern China*. Hong Kong: Ye Olde Printerie, 1954.

Phillips, Ralph W., Ray G. Johnson, and Raymond T. Moyer. *The Livestock of China*. Department of State Publication 2249, Far Eastern Series 9. Washington, DC: United States Government Printing Office, 1945.

"Pierce, James H. to L. F. Chen." March 2, 1949. A-3 Pierce Management 皮耳士經銷代理公司, AH 003-020400-0033.

"Plan for the Development of Coal and Mineral Resources of China." Pierce Management Incorporated, August 22, 1945. HST, Edwin A. Locke Jr. Files, Box 9, Folder: China Special.

"The President of the Chinese Executive Yuan and Acting Minister for Foreign Affairs (Chiang Kai-shek) to the American Ambassador (Gauss)." June 20, 1942. FRUS, China, 1942.

"Primer for the Chinese." *Investor's Reader* 5, no. 1 (July 18, 1945). Published by Merrill Lynch, Pierce, Fenner & Beane. FDR, AWPM, Box 13, Folder: International Training Administration, Inc.—China.

"Production of Alcohol in Free China." July 6, 1945. FDR, AWPM, Box 1, Liquid Fuels, Alcohol, General (Folder 1 of 2).

Proposals to the Export-Import Bank 1947 對進出口銀行提議, AH 003-020100-0515.

Pu, Min-Jen. "A Brief Report on the TVA." *National Reconstruction Journal* 4 (February 1944): 3–6.

Qian, Changzhao. "Correspondence with W. A. Haven in Arthur G. McKee and Company IC-4 阿瑟麥琪公司 函件," AH 003-020100-0210.

"Quantity and Model of Automobile in Free China." FDR, AWPM, Box 27, Folder: Transportation—Automotive and General.

Raj, Kapil. "Beyond Postcolonialism . . . and Postpositivism: Circulation and the Global History of Science." *Isis* 104, no. 2 (June 2013): 337–347.

"Ralph Strang to Andrew Kearney." April 26, 1945. FDR, AWPM, Box 15, Folder: Machine Tool Advisory Committee.

"Ralph Strang to Lester Bosch." September 26, 1945. FDR, AWPM, Box 30, Folder: Chinese Merit Awards—Personnel.

Rattenbury, Harold B. *Understanding China*. 2nd ed. London: Frederick Muller, 1943.

Reardon-Anderson, James. *The Study of Change: Chemistry in China, 1840–1949*. Cambridge: Cambridge University Press, 1991.

"Recruiting Technical Manpower." *CIE News Bulletin* 3, no. 1 (February 10, 1944).

"Regulations Governing the NRC Technical Group in USA." Technical Group of National Resource Commission, AH 003-020100-0429.

Ren, Naiqiang 任乃強. *Xikang tujing* 西康圖經. Taipei: Nantian chubanshe, 1987.

"Report of the Midwestern Conference of the Chinese National Reconstruction Forum." *National Reconstruction Journal* 5 (August 1944).

Reports of Yin-Kee Yuen S-406 袁炎基報告, AH 003-020600-1457.

Rodriguez, Andres. "Building the Nation, Serving the Frontier: Mobilizing and Reconstructing China's Borderlands during the War of Resistance (1937–1945)." *Modern Asian Studies* 2 (2011): 345–376.

Roland, Harold. "Alcohol Report." June 12, 1945. FDR, AWPM, Box 1, Folder: Liquid Fuels, Alcohol, General (Folder 1 of 2).

Rose, E. A. "The Sugar Industry at Taiwan—A Survey Report." E. A. Rose Incorporation (Sugar Industry) A Survey Report 羅斯公司臺灣糖業觀查報告, AH 003-020
200-0202.

———. "A Sugar Refining Industry for China." Memorandum to C. D. Shiah. E. A.
Rose Incorporation Reports of Taiwan Sugar Industry, 1947 一九四七年羅斯公司對臺
灣糖廠報告書, AH 003-020200-0222.

"Roy M. Jacobs to Andrew Kearney." March 24, 1945. FDR, AWPM, Box 15, Folder:
Machine Tool Advisory Committee.

"Rules for County and City Science Organs" 省市立科学馆规程. In 中华民国史档案资料
汇编·第五转第二遍 教育(2), edited by Zhongguo di er lishi dangan guan. Nanjing:
Fenghuang chubanshe, 2010.

Ryan, Jennifer, Lincoln C. Chen, and Tony Saich, eds. *Philanthropy for Health in China*.
Bloomington: Indiana University Press, 2014.

Schmalzer, Sigrid. "Breeding a Better China: Pigs, Practices and Place in a Chinese
County, 1929–1937." *Geographical Review* 92, no. 1 (January 2002).

———. *The People's Peking Man: Popular Science and Human Identity in Twentieth
Century China*. Chicago: University of Chicago Press, 2008.

Schneider, Laurence. *Biology and Revolution in Twentieth-Century China*. Lanham, MD:
Rowman & Littlefield, 2003.

Schoppa, R. Keith. *In a Sea of Bitterness*. Cambridge, MA: Harvard University Press,
2011.

"The Secretary of State to the Embassy in China." February 6, 1946. FRUS, The Far East:
China, 1946.

The Second China Education Yearbook 地二次中國教育年鑑. Nanjing: n.p., 1948.

Service, John Stewart. "Reminiscences of John Stewart Service." Berkeley Oral History.

Shang, Q., J. Lei, W. Jiang, H. Lu, and B. Liang. "Production of Tung Oil Biodiesel
and Variation of Fuel Properties during Storage." *Applied Biochemistry and Biotechnology* 168, no. 1 (September 2012): 106–115.

Shang, Q., W. Jiang, H. Lu, and B. Liang. "Properties of Tung Oil Biodiesel and Its
Blends with 0# Diesel." *Bioresource Technology* 1010, no. 2 (January 2010): 826–828.

Shen, Grace. "Murky Waters: Thoughts on Desire, Utility, and the 'Sea of Modern Science.'" *Isis* 98 (2007): 584–596.

Shen, Grace Yen. *Unearthing the Nation, Modern Geology and Nationalism in Republican China*. Chicago: University of Chicago Press, 2013.

Shen, Peiguang 沈培光. "Ma Baozhi and the Liuzhou School 馬保之与柳州牧校." Accessed August 31, 2016. http://blog.sina.com.cn/s/blog_63ad4fd70100mw21.html.

Shen, T. H. *Agricultural Resources of China*. Ithaca, NY: Cornell University Press,
1951.

———. *Autobiography of a Chinese Farmer's Servant*. Taipei: Linking Publishing, 1981.

———. "The Outlook of Application of Engineering Techniques to China's Agriculture." *Ta-Kung Newspaper* (Chungking), January 22, 1945.

Shi Yuanguang 师元光, ed. *The Pioneer in China Aviation, Wang Shizhuo* 中国航空事业
先驱王士倬. Beijing: Beijing fusheng yinshuachang, 2007.

Shih, Kuo-heng. *China Enters the Machine Age: A Study of Labor in Chinese War Industry*. Cambridge, MA: Harvard University Press, 1944.

Shih, Sam H. K. "Term Paper on Principles and General Techniques of Effective Teaching." William Homer Stacy Papers, Iowa State University Archives, Box 14, Folder 5.

Shire, A. C. "Abstract, Guide to the Industrialization of China." Washington, DC: Foreign Economic Administration, May 1945.

Sih, Paul K. T. *Nationalist China during the Sino-Japanese War, 1937–1945*. Hicksville, NY: Exposition Press, 1977.

"Sir H. Seymour to Mr. Eden." August 14, 1943. BDFA Vol. 7, Part 23.

"Sir H. Seymour to Mr. Eden." September 15, 1943. BDFA Vol. 7, Part 23.

"Sir H. Seymour to Mr. Eden." October 4, 1944. BDFA, Vol. 7, Part 26.

Soon, Wayne. *Global Medicine in China: A Diasporic History*. Stanford, CA: Stanford University Press, 2020.

Stacy, W. H. "Developing Voluntary Leadership in Extension Work." William Homer Stacy Papers, Iowa State University Archives, Box 14, Folder 5.

Stallings, Eugene. "Alcohol in China." February 26, 1945. FDR, AWPM, Box 1, Folder: Liquid Fuels, Alcohol—Miscellaneous Reports (Folder 1 of 2).

———. "Conference with Dr. H. Y. Chao, Chief of Alcohol Division, NRC." November 17, 1944. FDR, AWPM, Box 2, Liquid Fuels, Alcohol—Miscellaneous Reports (Folder 2 of 2).

———. "Report on Meeting on Alcohol." January 13, 1945. FDR, AWPM, Box 1, Folder: Liquid Fuels, Alcohol—Meetings and Conferences (Folder 2 of 2).

"Statement Concerning Administrative Costs of the Training Administration in Connection with FEA and ODT Programs from February, 1945 through October, 1946." 國際訓練經理社函件 Letters of International Training Administration Incorporation, AH 003-020600-1785.

Stein, Aurel. *Memoir on Maps of Chinese Turkistan and Kansu: From the Surveys Made during Sir Aurel Stein's Explorations, 1900–1, 1906–8, 1913–15*. Dehra Dun: Trigonometrical Survey Office, 1923.

Stross, Randall E. *The Stubborn Earth: American Agriculturalists on Chinese Soil 1898–1937*. Berkeley: University of California Press, 1986.

Sugar Industry Program Correspondences with Arthur G. Keller (Letters) 糖工業與美國合作計畫交涉函件, AH 003-020400-0025.

"Summary of Agreement between National Resources Commission of China and Arthur G. McKee & Company." October 24, 1945. McKee Co—Iron and Steel Works, 參琪公司—東北鋼鐵公司工作函件, AH 003-020400-0400.

"Summary of Conversation Held in Chungking on September 12, 1944, between Dr. Wong Wen-Hao Minister of Economic Affairs, and Mr. Nelson." FDR, Box 19, Folder: Post-war Industrialization Planning & Reconstruction Folder 2, September 44–45.

Sumner, John D. "Chinese Attitudes toward Postwar Economic Policy, Sept. 18, 1945." FDR, AWPB, Box 13, Interdepartmental Committee on Economic Policy toward China (Folder 1 of 2).

Surman, Jan. "Circulation, Globalization, Go-Betweens: Writing the Global History of 19th Century Science. Conversation with Kapil Raj." *Hioryka. Studia Metodologiczne* 46 (2016): 423–437.

Suttmeier, Richard P. *Research and Revolution*. Lanham, MD: Lexington Books, 1974.

Swen, W. Y. 孫文郁, and S. L. Chu 朱壽麟. *Production and Marketing of Wood Oil in Szechwan Province, China* 四川桐油之生產與運銷. Research Bulletin No. 7 (Chengtu Series). Nanjing: College of Agriculture and Forestry, Jinling University, February 1942.

"T. S. Lee to L. F. Chen." October 15, 1947. Letters and Cables of Taiwan Sugar Corporation 臺灣糖業公司來往函件及電報, AH 003-020400-0281.

"T. Y. Yen to Sandys Bao." November 8, 1948. Arthur G. McKee and Company IC-4 阿瑟參琪公司 函件, AH 003-020100-0210.

Taiwan Sugar-Bitting Incorporated, New York 美國毌替公司關於臺灣糖業報告, AH 003-020200-0448.

Tan, Ying Jia. *Recharging China in War and Revolution, 1882–1955*. Ithaca, NY: Cornell University Press, 2021.

Tang, Pei-Sung. *Green Thraldom: Essays of a Chinese Biologist*. London: Allen & Unwin, 1949.

———. "Science in China." *Science*, n.s., 91, no. 2358 (March 8, 1940): 38–39.

Tang, Tong B. *Science and Technology in China*. London: Longman, 1984.

Tao, Dinglai. "Brief History of Agricultural Engineering Development in China: In Memory of Mr. Zou Bingwen." *International Journal of Agricultural and Biological Engineering* (August 2008).

Taylor, Floyd. "I Attend a Chungking Exhibition: A Glimpse into Industrial China Tomorrow." *China at War* 12, no. 3 (March 1944): 64–68.

"Teaching United States Methods Builds Trade." Reprinted from the October *Red Barrel*, Magazine of the Coca-Cola Company, 1945. FDR, AWPM, Box 13, Folder: International Training Administration, Inc.—China.

"Technical Experts for China." October 7, 1944. National Archives (UK), F3137/16/10.

Teichman, Eric. *Journey to Turkistan*. London: Hodder and Stoughton, 1937.

"Telegram from Chiang Kaishek to the Ministry of Education" 蔣介石致教育部代电. November 10, 1942. In 中华民国史档案资料汇编，第五转第二遍 教育(2), edited by Zhongguo di er lishi dangan guan. Nanjing: Fenghuang chubanshe, 2010.

"Translation of a Text of Telegram from Ambassador Wei Tao Min, Washington, DC." February 4, 1945. Letters and Reports from War Production Board to Chairman Wen-Hao Wong 戰時生產局到翁文灝函件及電文, AH 003-020100-0589.

Trescott, Paul B. *Jingji Xue: The History of the Introduction of Western Economic Ideas into China, 1850–1950*. Hong Kong: Chinese University Press, 2007.

Tsai, Lung-Pao. "Innovation in Power Sources for Taiwan's Railways in the Period of US Aid (1950–1965)." In *The Development of Railway Technology in East Asia in Comparative Perspective*, edited by Minoru Sawai. Singapore: Springer, 2017.

Tseng, Chao-lun. "An Open Letter to British Scientists." *Acta Brevia Sinensia* 1 (January 1943).

Tsou, P. W. "China Must Have Agricultural Engineering." 1944. Accessed September 4, 2020. http://blog.sciencenet.cn/blog-39523-31245.html.

———. "Modernization of Chinese Agriculture." *Journal of Farm Economics* 28, no. 3 (August 1946): 773–790.

Tsou, P. W., and C. W. Chang. "A Program for Postwar Agricultural Reconstruction in China." FDR, AWPM, Box 1, Agriculture and Forestry (Folder 1 of 2).

Tsou, Ping-wen. "A Fundamental Problem of Economic Reconstruction in China." *National Reconstruction* 5 (July 1944).

Tsou, Tang. *America's Failure in China, 1941–1950*. Chicago: University of Chicago Press, 1963.

Tuchman, Barbara W. *Stilwell and the American Experience in China, 1911–1945*. New York: Macmillan, 1971.

Tung, William L. *V. K. Wellington Koo and China's Wartime Diplomacy*. New York: St. John's University, 1977.

U.S. Senate Committee on Foreign Relations. Report no. 1026 "Aid to China." Washington, DC: Government Printing Office, 1948.

United States Congress, Committee on Foreign Affairs. *Review of the Mutual Security Program Development and Implementation*. Washington, DC: United States Government Printing Office, 1958.

United States Department of State. *The China White Paper, August 1949*. Stanford, CA: Stanford University Press, 1967.

United States Economic Cooperation Administration. *Mission to China, U.S. Economic Assistance to Formosa, 1 January to 31 December 1950*. Washington, DC: United States Economic Cooperation Administration, 1951.

"Universal Trading Corporation." *Fortune* 22, no. 6 (December 1940).

van de Ven, Hans. *War and Nationalism in China, 1925–1945*. London: Routledge, 2003.

Vouchers of February 1952 資源委員會一九五二年二月經手帳單(一), AH 003-020300-0620.

Wakeman, Frederic Jr., and Richard Louis Edmonds. *Reappraising Republican China*. Oxford: Oxford University Press, 2000.

Walker, M. N. Memorandum to W. T. Stanton, "Alcohol Report of Mr. Harold Roland." June 28, 1945. FDR, AWPM, Box 1, Folder: Liquid Fuels, Alcohol, General (Folder 1 of 2).

"Walter G. Whitman to James Jacobson." October 24, 1944. FDR, AWPM, Box 15, Miscellaneous (Folder 2 of 4).

Wan, Charles. "Work of National Highways Bureau." *China at War* 12, no. 5 (May 1944).

Wang, Shou-chin. "Alcohol Factories." October 7, 1944. FDR, AWPM, Box 2, Folder: Liquid Fuels, Alcohol—Miscellaneous Reports (Folder 2 of 2).

Wang, Zuoyue. "The Chinese Developmental State during the Cold War: The Making of the 1956 Twelve-Year Science and Technology Plan." *History and Technology* 31, no. 3 (2015): 180–205.

Weller, J. Marvin. *Caravan across China: An American Geologist Explores the Northwest, 1937–1938*. San Francisco: March Hare, 1984.

———. "Outline of Chinese Geology." *AAPG Bulletin* 28, no. 10 (October 1944): 1417–1429.

"Weng Wenhao to Donald Nelson." March 30, 1945. FDR, AWPM, Box 27, Folder: Transportation—Automotive and General.

"Weng Wenhao to Locke." December 24, 1945. HST Library, Edwin A. Locke Jr. File, Box 2, Folder: Mission to China—3rd Trip: China (Reports—Supreme Economic Council).

"Weng Wenhao to Howard Coonley." February 20, 1945. Letters and Reports from War Production Board to Chairman Wen-Hao Wong 戰時生產局到翁文灝函件及電文, AH 003-020100-0589.

Whitney, Le Roy. "Transportation in China." November 7, 1944. FDR, AWPM, Box 27, Folder: Transportation—Automotive and General.

Willauer, Whiting, and J. Franklin Ray. "As Paraphrased in Foreign Economic Administration, China Advisory Committee, Meeting of March 13, 1945." FDR, AWPM, Box 8, Folder: Foreign Economic Aid.

"William A. Haven Curriculum Vita." Arthur G. McKee and Company IC-4 阿瑟參琪公司 函件, AH 003-020100-0210.

"William L. C. Hui to Yuen Yin-kee." October 31, 1947, and other documents. Tung-Yuan Chen and Yin-Kee Yuen Introduction Letters 陳東元及袁銀基推介函件, AH 003-020600-2844.

Wong, Kevin Scott. *Americans First: Chinese-Americans and the Second World War.* Cambridge, MA: Harvard University Press, 2005.

Wu Daguan. "Wu Daguan's Oral Autobiography: My Chinese Heart" 吴大观口述自转: 我的中国心. Accessed August 23, 2016. http://www.doc88.com/p-990191936223.html.

Wu, K. C. "Memorandum on China's Urgent Requirements of Industrial Equipment and Materials." September 15, 1944. FDR, AWPM, Box 10, Folder: Industry, General Reports (Folder 1 of 3).

"Wu, W. H. Memorandum to L. F. Chen." September 16, 1943. Letters of Professor Paul B. Eaton, AH 003-020600-1760.

Xiang, Xiufang 尚季芳. "A Review on the Northwestern Explorers and Their Works in the Period of National Government" 国民政府时期的西北考察家及其著作述评. *Zhongguo bianjiang shi di yanjiu* 13, no. 3 (September 2003): 106–113.

Xie, Jiarong 謝家榮. "Our Petroleum Prospect" 中国之石油. *Dili xuebao* 2, no. 1 (1935).

Xue Yi 薛毅. "The Thirty–One Scholars of the War of Resistance" 抗战时期的三一学士. *Kangri zhanzheng yanjiu* 2 (2003): 87–107.

"Y. C. Young to O. R. Johnson." August 30, 1945. FDR, Box 10, Folder: Formosa.

Yang, J. J., Y. Li, and W. Pan. "Biodiesel Production from a Novel Raw Material Tung Oil." *Advanced Materials Research* 608–609 (2013): 406–410.

Yang, K. H. "Raw Materials for Alcohol." November 20, 1944. FDR, AWPM, Box 1, Folder: Liquid Fuels, Alcohol, General (Folder 2 of 2).

Yang, Shu-chia. "Fundamental Problems of Chinese Agriculture." *National Reconstruction* 5 (July 1944).

Yang, Y. C. "Chinese Education, Past and Present." Southern University Conference, 1944: Proceedings Constitution and By-Laws Addresses Reports, Atlanta, Georgia, April 12–13, 1944.

Yeh, Wen-hsin, ed. *Wartime Shanghai.* New York: Routledge, 1998.

Yen, Paul H. "Victory Alcohol Plant and the Alcohol Industry in Szechwan." November 28, 1944. FDR, AWPM, Box 2, Folder: Liquid Fuels, Alcohol Plants (other than those on which inspections have been reported).

Yin, Liangwu. "The Long Quest for Greatness: China's Decision to Launch the Three Gorges Project." PhD diss., Washington University in St. Louis, 1996.

Young, Arthur N. *China's Wartime Finance and Inflation, 1937–1945.* Cambridge, MA: Harvard University Press, 1965.

Yu, Kuo-ping. "Tung Oil Production and Its World Market." MA thesis, University of Southern California, 1949.

Yui Shin-min. "China's Tung Oil Export and Its Future Prospects." *National Reconstruction Journal* 5 (1944): 99–117.

"Yun, C. Memorandum to W. S. Finlay Jr." June 12, 1947. Letters between National Resources Commission Technical Office in United States of America and the J. G. White Engineering Corporation 資源委員會朱美技術團與懷特工程公司往來函件, AH 003-020400-0237.

Zanasi, Margherita. *Saving the Nation: Economic Modernity in Republican China.* Chicago: University of Chicago Press, 2006.

Zhan, Yongfeng, Wang Hongbo, and Deng Hui. "The General View about the China Institute of Geography during the Republic of China Period" 民國時期中國地理研究所鈎沉. *Dili yanjiu* 33, no. 9 (September 2014): 1768–1777.

Zhang, Songyin 張松荫. "Livestock of Southwest Gansu" 甘肅西南之畜牧. *Dili xuebao* 9 (1942).

Zhao, Wei, Liu Laping, Yang Jian, Ma Yangmin, Zhou Le. "Preparation of Biodiesel Oil from Tong Oil" 桐油转化生物柴油的研究. *Journal of Northwest A&F University* 35, no. 11 (November 2007): 176–180.

Zheng, Huixin 郑会欣. *Research on the Nationalist Government's Wartime Control of Economy and Trade* 国民政府战时统制经济与贸易研究. Shanghai: Shanghai shehui kexue yuan, 2009.

Zhou Lisan 周立三, Hou Xuedao 侯學燾, and Chen Siqiao 陳泗橋. *Economical Atlas of Sichuan* 四川經濟地圖集. Beibei: China Institute of Geography, 1946.

Zhu, Pingchao. *Wartime Culture in Guilin, 1938–1944: A City at War.* Lanham, MD: Lexington Books, 2015.

Index

Page numbers for figures and tables are in italics.

Harvard East Asian Monographs
(most recent titles)